Zillions of Practice Problems

Pre-Algebra 1 with Biology

Zillions of Practice Problems
Pre-Algebra 1 with Biology

Stanley F. Schmidt, Ph.D.

Polka Dot Publishing

© 2017 Stanley F. Schmidt
All rights reserved.

ISBN: 978-1-937032-59-3

Printed and bound in the United States of America

Polka Dot Publishing   Reno, Nevada

Order Life of Fred Books from:
JOY Center of Learning
http://www.LifeofFredMath.com

Questions or comments? Email the author at lifeoffred@yahoo.com

First printing

Zillions of Practice Problems Pre-Algebra 1 with Biology was illustrated by the author with additional clip art furnished under license from Nova Development Corporation, which holds the copyright to that art.

for Goodness' sake

or as J.S. Bach—who was never noted for his plain English—often expressed it:

Ad Majorem Dei Gloriam
(to the greater glory of God)

If you happen to spot an error that the author, the publisher, and the printer missed, please let us know with an email to: lifeoffred@yahoo.com

 As a reward, we'll email back to you a list of all the corrections that readers have reported.

How This Book Is Organized

Life of Fred: Pre-Algebra 1 with Biology has 45 chapters before the entre nous extra chapter. So does this book.

As you work through each chapter in *Life of Fred: Pre-Algebra 1 with Biology* you can do the problems in the corresponding chapter in this book.

Each chapter in this book is divided into two parts.

☆ The first part takes each topic and offers a zillion problems.

☆ The second part is called the Mixed Bag. It consists of a variety of problems from the chapter and review problems from the beginning of the book up to that point.

Please write down your answers before turning to the back of the book to look at my answers. If you just read the questions and then read my answers you will learn very little. As Thomas Fuller (in about 1700) used to say,

He that will have the Kernel must crack the Shell.

Chapter One
Living Things

First part: Problems from this chapter
with a small look back to arithmetic

114. $\dfrac{5}{6} + \dfrac{3}{4}$

218. $\dfrac{5}{6} \times \dfrac{3}{4}$

321. $\dfrac{5}{6} \div \dfrac{3}{4}$

412. Which is smaller: $\dfrac{5}{6}$ or $\dfrac{3}{4}$?

560. Convert $\dfrac{5}{6}$ to a percent.

658. 3 is what percent of 4?

728. What is 15% more than 60?

A small thought: If you had trouble with these problems, it might be a good idea to take a week off and spend it reviewing arithmetic.

arithmetic ↔ pre-algebra ↔ algebra

Without knowing arithmetic, pre-algebra is much harder than it needs to be. Reviewing arithmetic will make your life happier.

Chapter Two
What Is Life?

First part: Problems from this chapter

150. Right now, you could divide the entire physical universe into two sets: those things that are not alive and those things that are alive.

The **union** of those two sets is equal to every physical thing in the universe.

Those two sets are **disjoint** because those two sets do not have anything in common. You are either not alive or you are alive.

Suppose I divide the entire physical universe into these two sets: Set A = everything that weighs less than 4 kilograms and Set B = everything that weighs more than 3 kilograms.
Is the union of A and B equal to the entire physical universe?
Are A and B disjoint?

272. Suppose set E = everything that you have ever seen or touched. Let set F = everything that your best friend, Jan, has ever seen or touched.
Does the union of E and F equal the entire physical universe?
Are E and F disjoint?

454. Every sticker in my sticker collection was either given to me or I bought it at the store. If set G = those stickers that have been given to me and if set H = those stickers I bought at the store, then is the union of G and H equal to my sticker collection? Are G and H disjoint?

525. You and your best friend Jan head off to Ivy's Ice Cream. Ivy has 1,000 ice cream flavors. You like 600 of Ivy's flavors. Jan like 550 of Ivy's flavors.
Let set K = the flavors you like.
Let set L = the flavors Jan likes.
Must the union of K and L (known as K ∪ L) equal all of Ivy's flavors?
Must K and L be disjoint?

Chapter Two   What Is Life?

Second part: the 𝕄ixed 𝔹ag: a variety of problems from this chapter and previous chapters

659. Jan looked at the 1,000 flavors. One of them was anchovy–lamb with a ribbon of yam. Jan asked Ivy how she dreamed up all those 1,000 flavors of ice cream.

Ivy explained that she researched every ice cream store in the world and used every flavor she found.

Does this mean that the list of flavors that Ralph's Three Flavor ice cream store and the list of flavors of Ivy's Ice Cream are not disjoint?

727. Jan asked, "Do many people buy the anchovy-lamb with a ribbon of yam?"

Before Ivy could answer, Jan repeated:
aNChoVY-LaMB With a RiBBoN oF YaM
   aNChoVY-LaMB With a RiBBoN oF YaM
      aNChoVY-LaMB With a RiBBoN oF YaM
         aNChoVY-LaMB With a RiBBoN oF YaM
            aNChoVY-LaMB With a RiBBoN oF YaM

Jan broke out into song. She danced around in Ivy's Ice Cream store. It was like watching a Hollywood musical, except that this was really happening. Jan is a very happy person, but she can sometimes be a little embarrassing to be around.

In the previous year only 4 customers had ordered anchovy-lamb with a ribbon of yam. Today, with Jan's singing and dancing, 7 people got up the courage to order that flavor. Seven is what percent more than 4?

829. Ivy couldn't believe the effect that Jan's singing and dancing had on ice cream orders.

Ivy had the largest ice cream store in the world. She carried every flavor of ice cream known to mankind. And she was going broke.

As everyone in business knows, your profit equals your income minus your expenses. $P = I - E$

Does a business owner want: $I > E$, $I = E$, or $I < E$?

922. You and Jan head off to Stanthony's PieOne pizza place for lunch. You have a pepperoni pizza, a salad, a scoop of anchovy-lamb with a ribbon of yam on an ice cream cone and a half dozen other items. Name the one thing that you had that wasn't once alive.

Chapter Three
Circular Definitions

First part: Problems from this chapter

128. During lunch you noticed that Jan wiggled a lot. "As a theater arts* major," Jan explained, "I have to learn a lot of dancing. My body is always moving."

One-third of the way through the meal she had already eaten 2,877 Calories. At that rate, how many Calories would she have for lunch?

251. You were two-fifths of the way through your lunch and had eaten 347 Calories. At that rate, how many Calories would you be having for lunch?

You don't eat as much as Jan does.

410. Your lunch bill was $4.89. Jan's was $11.22. Jan's bill was what percent greater than yours? Round your answer to the nearest percent.

558. Over lunch Jan explained to you all the things a theater arts major had to learn. There was dancing. There was singing. And—surprise!—there was acting.

And there were many kinds dancing: ballet, jazz, tap, and so on.

That morning Jan had done 2½ hours of tap dancing. According to Prof. Eldwood's *Tap Dancing through Life*, 1843, a person spends 892 Calories in tap dancing for 40 minutes (40 minutes = ⅔ hour).

How many Calories had Jan spent tap dancing this morning? (Use conversion factors.)

---

* Most theater arts majors spell it *theatre arts*. That makes it look more fancy. Do you ever see American mathematicians write about the *centre* of a circle? *Centre* is the British spelling of *center*.

Chapter Three   Circular Definitions

Second part: the 𝔐ixed 𝔅ag: a variety of problems from this chapter and previous chapters

664. Jan did a lot of talking during lunch. It is fun to listen to what the world looks like according to Jan.

Jan: "The world is divided into two groups—theater arts majors and the audience."

✷ Are those two sets disjoint?

✷ Is the union of those two sets equal to the set of all the people in the world?

726. Exercise makes a body grow stronger. You can do 6 pushups. Jan can do 100% more pushups than you. How many can Jan do?

858. You can run 8 mph (miles per hour). Jan can run 26.8% faster than you. How fast can Jan run? (Round your answer to the nearest tenth of a mile per hour.)

860. Jan tells you how wonderful it is to be an actor.

You get to dress up. (You think: But you don't get to keep the clothes.)

You get to sign autographs. (You think: The only autographs that mathematicians deal with are autographs on checks that are made out to the mathematician. The average mathematician makes a lot more money than the average actor.)

You get to have an agent. (You think: Mathematicians don't need agents who take a percentage of your earnings. An actor has a job that often only lasts for weeks or months. A mathematician's job often lasts for years.)

Jan breaks into song—"Hi-diddle-dee-dee, an actor's life for me"—and you think of when Pinocchio, the puppet-boy, was tempted to go to Stromboli's Caravan to become a actor in the 1940 movie. It didn't turn out well. The lunch bill comes. Jan is "a little short of funds right now." Guess who pays for the lunch.

Chapter Four
Starting a Garden

First part: Problems from this chapter

204. This last summer Jan was in a show. It went three weeks before it closed. Each night from 6 p.m. to 10 p.m. for 21 nights. "I made $25 each night," Jan told you. How much did Jan make in those three weeks?

207. Out of Jan's paycheck the government took 20% in taxes. How much was left for the three weeks' of work?

369. Jan had worked for four hours per night for 21 nights and had made $420. Using conversion factors determine how much this actress made per hour.

Start with $\dfrac{\$420}{\text{the whole job}}$

459. Each night Jan arrived at the theater at 6 p.m. The first 45 minutes were spent putting on stage makeup and getting into costume. After the performance itself, it took 30 minutes to take off the makeup and get out of the costume. We already know how much Jan was paid for those four hours. If you have ever done acting, you know that there were hours that Jan wasn't paid for. What were they?

660. You are almost finished with your lunch with Jan at Stanthony's PieOne pizza. Ivy rushes in. "I'm so glad I found you!" she exclaims. "Ever since you sang and danced at my ice cream store, I've had zillions of customers. The news got out. They came to see you. And they bought ice cream!"

    Jan said, "I'm so happy to hear that."

    Ivy shouts, "You don't understand. I'll pay you $200 an hour to do that aNChoVY-LaMB WiTh a RiBBoN oF YaM song and dance. You can set your hours—whenever you want."

    Jan winks at you as they leave to head to Ivy's Ice Cream.

    You think: *$200 per hour, 40 hours per week, 50 weeks per year.* What would be Jan's annual income?

Chapter Four   Starting a Garden

Second part: the 𝔐ixed 𝔅ag: a variety of problems from this chapter and previous chapters

723. That afternoon the school newspaper broke the news.

THE KITTEN Caboodle

The Official Campus Newspaper of KITTENS University   Wednesday 3:20 p.m. Edition 10¢

exclusive
Jan Headlines at Ivy's

KANSAS: All of Kansas is talking about the new song-and-dance act at Ivy's Ice Cream store.
    Thousands have flocked to witness this ground-breaking advance in entertainment.
    Movie theaters have closed their doors.
    Classrooms are empty.
    Football games have been canceled.

—Ivy
"We also have ice cream for sale."

Everyone is talking about 𝕒ℕ𝒸𝒽𝑜𝕍𝕐▪𝕃𝕒𝕄𝔹 𝕎𝕚𝕥𝕙 𝕒 ℝ𝕚𝔹𝔹𝑜ℕ 𝑜𝔽 𝕐𝕒𝕄.

Ivy normally gets her hair done for $30. After hearing about Ivy's success, her hairdresser increased Ivy's bill to $35. What percentage increase was that?

830. In order to get into Ivy's and see Jan perform, you have to buy 6 cartons of 𝕒ℕ𝒸𝒽𝑜𝕍𝕐▪𝕃𝕒𝕄𝔹 𝕎𝕚𝕥𝕙 𝕒 ℝ𝕚𝔹𝔹𝑜ℕ 𝑜𝔽 𝕐𝕒𝕄 and a $5 plastic spoon. The whole package costs $24.62.
    How much does a carton of 𝕒ℕ𝒸𝒽𝑜𝕍𝕐▪𝕃𝕒𝕄𝔹 𝕎𝕚𝕥𝕙 𝕒 ℝ𝕚𝔹𝔹𝑜ℕ 𝑜𝔽 𝕐𝕒𝕄 cost?

Here are some guidelines for solving "word problems" (as they are called in algebra).
① Let x equal the thing you are trying to find out. This means reading the problem and understanding the English. Often, looking for the question mark will be a good place to learn what is wanted.
② Write Then xxx equals yyy statements that depend on the Let x = statement you wrote in line ①.
③ After you have written several Then xxx equals yyy statements the equation will almost write itself.

Chapter Five
Seeds and Water

First part: Problems from this chapter

206. Solve $4x + 7 = 29.8$
First subtract 7 from both sides. Then divide both sides by 4.

220. A carton of 𝗮𝗡𝗖𝗵𝗼𝗩𝗬-𝗟𝗮𝗠𝗕 𝗪𝗶𝘁𝗵 𝗮 𝗥𝗶𝗕𝗕𝗼𝗡 𝗼𝗙 𝗬𝗮𝗠 ice cream measures $2.2" \times 3" \times 4.8"$. (" means inches)
   What is the volume of this box?

277. The density of 𝗮𝗡𝗖𝗵𝗼𝗩𝗬-𝗟𝗮𝗠𝗕 𝗪𝗶𝘁𝗵 𝗮 𝗥𝗶𝗕𝗕𝗼𝗡 𝗼𝗙 𝗬𝗮𝗠 ice cream is 0.09 pounds per cubic inch. How many pounds does a carton of that ice cream weigh? Round your answer to the nearest pound.

A density of 0.09 pounds per cubic inch means that the conversion factor will be either
$\frac{0.09 \text{ pounds}}{1 \text{ cubic inch}}$ or it will be $\frac{1 \text{ cubic inch}}{0.09 \text{ pounds}}$
You know the volume from the previous problem.

456. As you might have guessed, 𝗮𝗡𝗖𝗵𝗼𝗩𝗬-𝗟𝗮𝗠𝗕 𝗪𝗶𝘁𝗵 𝗮 𝗥𝗶𝗕𝗕𝗼𝗡 𝗼𝗙 𝗬𝗮𝗠 ice cream smells really 𝗕𝗔𝗗. People would happily trade 7 cartons of 𝗮𝗡𝗖𝗵𝗼𝗩𝗬-𝗟𝗮𝗠𝗕 𝗪𝗶𝘁𝗵 𝗮 𝗥𝗶𝗕𝗕𝗼𝗡 𝗼𝗙 𝗬𝗮𝗠 ice cream for 3 cartons of strawberry ice cream. If you had 623 cartons of the smelly stuff, how many cartons of strawberry ice cream could you trade it for?

The conversion factor will be either $\frac{3 \text{ strawberry}}{7 \text{ anchovy}}$ or $\frac{7 \text{ anchovy}}{3 \text{ strawberry}}$

Wait! Stop for a moment. I, your reader, have a question. I understand why people would trade their anchovy-lamb with a ribbon of yam ice cream for some strawberry ice cream. But why would anyone want to give up their strawberry ice cream for the smelly stuff?

In economics we know that for a trade to happen, both sides must be happy with what they are getting.

Cat owners have found that their cats really love 𝗮𝗡𝗖𝗵𝗼𝗩𝗬-𝗟𝗮𝗠𝗕 𝗪𝗶𝘁𝗵 𝗮 𝗥𝗶𝗕𝗕𝗼𝗡 𝗼𝗙 𝗬𝗮𝗠 ice cream. It's cheap cat food.

Chapter Five   Seeds and Water

Second part: the 𝔐ixed 𝔅ag: a variety of problems from this chapter and previous chapters

559. A carton of Ivy's strawberry ice cream costs 42¢. If a cat owner buys 3 cartons of strawberry ice cream and then trades it for 7 cartons of 𝖆𝖓𝖈𝖍𝖔𝖛𝖞-𝖑𝖆𝖒𝖇 𝖜𝖎𝖙𝖍 𝖆 𝖗𝖎𝖇𝖇𝖔𝖓 𝖔𝖋 𝖞𝖆𝖒 ice cream, they have very cheap cat food that their cats adore. How much would a carton of this cat food cost the cat owner?

We'll use conversion factors. Start with one carton of 𝖆𝖓𝖈𝖍𝖔𝖛𝖞-𝖑𝖆𝖒𝖇 𝖜𝖎𝖙𝖍 𝖆 𝖗𝖎𝖇𝖇𝖔𝖓 𝖔𝖋 𝖞𝖆𝖒 ice cream.

The first conversion is from anchovy to strawberry. That conversion factor will be either $\frac{3 \text{ strawberry}}{7 \text{ anchovy}}$ or $\frac{7 \text{ anchovy}}{3 \text{ strawberry}}$

The second conversion factor will be from cartons of strawberry to price. It will be either $\frac{\text{one carton of strawberry}}{42¢}$ or $\frac{42¢}{\text{one strawberry}}$

662.

Ivy's favorite book is Prof. Eldwood's *Getting Rich through Ice Cream*, 1846.

Ivy's annual expenses are:
Building and utilities $42,000.
Cost of ice cream $377,000.
Paying Jan to sing and dance $400,000.

Ivy's annual income is:
        $2,500,000.

What is Ivy's annual profit?

859. When they get a high income, some people feel the *need* to spend it. There are many stories of boxers and football players earning millions of dollars each year. They buy expensive houses and fancy cars to show off. When they get really old (like 35 years old) their income stops. And in a couple of years, they are broke.

Ivy is no dummy. She invests most of her profits rather than spending them. If she invests a million dollars and it earns 5% per year, how much will she have in a year?

Chapter Six
Cones

First part: Problems from this chapter

123. Jan bought 5 new shirts and a $6 hair ribbon for her new dancing job at Ivy's Ice Cream. The cost was $73. How much was each shirt? (The shirts were all the same price.)
You begin by letting x = the cost of one shirt. (That is what you are trying to find out.) Then 5x = and so on.

276. Jan is paid $200 per hour for her singing and dancing. The $73 that she just spent on shirts and a ribbon is what percent of an hour's work?

326. How many minutes of singing and dancing would it take to buy that stuff?

430. Ivy bought a giant ice cream cone to place outside her building. The cone is 50 feet tall and 20 feet across at the top. What is its volume? (Round your answer to the nearest cubic foot.)

561. Ivy's annual income this year is $2,500,000. She anticipates that it will grow by 4% per year in the future. How much will her income be two years from now?

663. Ivy's giant ice cream cone (see two problems ago) was shipped to the front of her store in a giant cylinder to protect it from damage.

You know from the current chapter that $V_{cone} = \frac{1}{3}\pi r^2 h$.

Make a guess . . . the cone occupies what fraction of the volume of the cylinder that encloses it?

Chapter Six   Cones

Second part: the 𝔐ixed 𝔅ag: a variety of problems from this chapter and previous chapters

731. Ivy wasn't sure what to do with that cylinder (50 feet tall and with a diameter of 20 feet). She decided to fill it with "leftover" pennies that she received when selling ice cream.*

    There are approximately 50,000 pennies in a cubic foot. (← That would make a nice conversion factor!)

    Roughly, how many pennies will that cylinder hold?

This is a two-step problem. First, find the volume of the cylinder and then use a conversion factor.

855. Solve $8x + 5 = 31$

861. Which is larger 3^2 or 2^3?

950.

If you are thinking of ordering some 𝒶𝓃𝒸𝒽𝑜𝓋𝓎-𝓁𝒶𝓂𝒷 𝓌𝒾𝓉𝒽 𝒶 𝓇𝒾𝒷𝒷𝑜𝓃 𝑜𝒻 𝓎𝒶𝓂 ice cream, it would be nice to learn a little bit about anchovies.

    The unimportant facts are . . .

✢ They are a small green fish with touches of blue.

✢ They are caught, gutted, and stored for a while in brine,** which does two things: It turns them dark gray and it increases their flavor.

✢ You can buy them packed in salt or packed in oil or smashed*** into a paste.

    The important fact is . . .

✢ They have a strong flavor—powerfully strong. I like them on pizzas. Most of my friends can't stand them.

    Your question: Are you more likely to find anchovy ice cream or an eagle who can do calculus?

---

* The nickels, dimes, quarters, dollar bills, five-dollar bills, ten-dollar bills, twenty-dollar bills, fifty-dollar bills, and hundred-dollar bills she sent to the bank.

** Brine = strong salt water

*** The official cooking word for "smashed into a paste" is puréed.

Chapter Seven
Life Cycles

First part: Problems from this chapter

129. Ivy's Ice Cream offers 1,000 flavors. 57% of them are not very popular. (They are ordered less than one time per month.)
  How many flavors are not very popular?

217. Seventeen of the ice cream flavors are popular. (They are ordered more than 40 times per month.) What percent of the flavors are popular?

320. You went to see Jan sing and dance. It had been two weeks since she got this job with Ivy's Ice Cream. She looked tired. Eight hours of singing and dancing each day were starting to wear her out.
  The aNchoVY-LaMB WiTh a RiBBoN oF YaM song was the only song she was allowed to sing. Sometimes she would sing it fast.* Sometimes she would sing it slowly. Sometimes, loudly. Sometimes, softly.
  During her five-minute break she came to talk with you. She drank a two-dollar bottle of cola and washed down 7 purple "throat pills." That's what she called them. The pills and the cola cost $49.46. How much did each pill cost?

462. Before Jan started performing, Ivy was filling 60 ice cream orders each day. After a week, Ivy will filling 230% more orders each day. How many orders is that?

---

* *Fast* is both an adjective and an adverb.
  Adjectives modify people, places and things (nouns). The fast runner. The fast car. Some beginning writers tend to overuse adjectives. They might write, "The *loud* clash of cymbals and the *dancing* feet of the *spinning* waltzers...."
  Adverbs modify verbs (He drove *slowly*), adjectives (It was *bright* red), or other adverbs (Jan sang *quite* slowly).

Chapter Seven   Life Cycles

Second part: the 𝔐ixed 𝔅ag: a variety of problems from this chapter and previous chapters

527. The world can be very fickle.* The toys and video games that were popular last year are "old stuff" this year.

After two weeks the excitement over the 𝗮𝗡𝗰𝗵𝗼𝗩𝗬-𝗟𝗮𝗠𝗕 𝗪𝗶𝘁𝗵 𝗮 𝗥𝗶𝗕𝗕𝗼𝗡 𝗼𝗙 𝗬𝗮𝗠 song had died. The 198 ice cream orders per day had dropped by 97.5%. Approximately how many orders were now being received?

665. Jan realized that there were fewer orders than when she first began to sing and dance. This was scary. Was her singing driving people away from Ivy's?

At 5 p.m. Ivy walked toward Jan.

Termination
   Dismissal
      Getting the sack
         Being let go
            Receiving walking papers
               Fired

Ivy didn't have to explain. She didn't use nasty words like "You're toast" or "You stink." Ivy gave her $1,600 for her days' work. They hugged. Jan headed back to her apartment.

Happily, Jan hadn't bought a mansion. She had been thinking of doing that given her anticipated $400,000/year income. But Jan knew that being in the theatrical arts world has many more ups and downs than working as a cashier in a grocery store.

She had 63 boxes of macaroni-and-cheese. Eating 3 boxes per day, how long would that last? For the practice, please use a conversion factor.

---

* Not sure what *fickle* means? You have three choices: ❶ Stay ignorant, ❷ You do own a dictionary, don't you? or ❸ Find out its meaning from the context in which it is used.

Regarding ❶ . . . having a decent-sized vocabulary can have a lot of advantages. One example: When I entered college a zillion years ago, each incoming freshman was given an English test to determine whether they would be allowed into college English. One of the questions was to define the word *saturnine*. I confused it with *saturnalia*, which means big partying like on New Year's eve. Luckily, I knew enough of the other words to get admitted into the college English course.

Chapter Eight
Tooth Brushing

First part: Problems from this chapter

119. Calculators are neat. They can take away a lot of work when you have to divide 41,964 by 78. That beats doing

$$\begin{array}{r} 538 \\ 78\overline{)41964} \\ \underline{390} \\ 296 \\ \underline{234} \\ 624 \\ \underline{624} \\ 0 \end{array}$$

But they aren't much help when you are adding fractions and you have to find a least common denominator (also known as an LCD).

The LCD for $\frac{3}{4}$ and $\frac{5}{6}$ is 12.

Find the LCD for each of these:

A) $\frac{5}{6}$ and $\frac{3}{8}$

B) $\frac{3}{10}$ and $\frac{1}{4}$

C) $\frac{7}{10}$ and $\frac{2}{15}$

254. Things get much more exciting when you have several fractions that you want to add.
Find the LCD of $\frac{1}{2}$ and $\frac{1}{3}$ and $\frac{1}{4}$ and $\frac{1}{5}$

302. Jan's 63 boxes of macaroni and cheese plus a $1.13 plastic bowl cost a total of $32. How much does a box of macaroni and cheese cost?

458. Living on three boxes of macaroni and cheese each day was certain to have an effect on Jan's weight. It is the kind of diet that you don't want to be on. If she started at 118 pounds and dropped to 100 pounds, what percent would she have lost? Give your answer to the nearest percent.

22

Chapter Eight   Tooth Brushing

Second part: the 𝔐ixed 𝔅ag: a variety of problems from this chapter and previous chapters

655. Jan needed some cash and she needed it now. She called her favorite director and asked if there were any acting parts available. The director said that they were putting on Shakespeare's *A Midsummer Night's Dream*.*

"Great!" Jan exclaimed.

"There is only one part that we haven't cast yet. Do you remember the donkey in that play?"

"Of course," Jan replied. "The donkey scenes are some of the funniest parts of that play."

"The part we haven't cast yet is the back end of the donkey. It takes two actors to fill the costume. Are you interested?"

Jan took one look at the remaining boxes of macaroni and cheese and said, "Yes! When do we start rehearsals?"

"At 6 p.m."

"I'll be there."

It's now 4:15 p.m. How long does Jan have to get there?

I, your reader, have a question before you go any further.** Jan is now desperate for money. She's living on macaroni and cheese. What happened to the $1,600 she was making each day for a couple of weeks?

It's true that she didn't go out and buy a mansion. But she did go out and buy lots of doodads, thingamabobs, and bibelots.*** She spent her money as fast as it came in with the expectation that the money faucet

---

* I have seen three different filmed versions of this play. My favorite is the 1935 version in which Mickey Rooney plays Puck. I love the way he laughs and jumps around in that movie.

** Use *farther* when dealing with actual physical distance. In other cases, use *further*.

*** A *bibelot* (pronounced BIB-low or, if you want to sound more French, BEEB-beh-LOW) is some small object like a statue of a seven-fingered monkey. A bibelot is usually pretty or rare or interesting in some way.

Chapter Eight   Tooth Brushing

would always be turned on full force. The ten-dollar bibelots she bought wouldn't fetch 50¢ at a garage sale.

Many Americans spend that way. They live paycheck to paycheck, and it doesn't matter whether they are making a lot of money or very little. The act of saving or investing never happens. They even spend money they don't have by running up the balances on their credit cards.

Jan spent her last $1,600 trying to pay down her credit cards. That's why she's living on macaroni and cheese.

Thank you.

729. Jan had charged an equal amount on her seven credit cards (Disaster Card, Pleasa Card, American Distress, Uncover Card, etc.) and she had borrowed $100 from you. The total came to $21,800. How much did she owe on each credit card?

826. At 6 p.m. Jan arrived at the rehearsal.

The Cast List—A Midsummer Night's Dream

SPEAKING PARTS

    Lysander... Alex
    Demetrius.. Bob
    Hermia.. Cassie
    Helena.. Diana
    Oberon, King of Fairies................................ Edgar
    Titania, Queen of Fairies............................... Fay
    Puck, Oberon's jester........................... Mickey R.
    Front end of donkey................................. Gordon

NON-SPEAKING PARTS

    Donkey butt... Jan

These two sets—the speaking parts and the non-speaking parts—are all the people acting in the play. The union of these two sets is equal to all the actors in the play. Are the two sets disjoint?

901. Let P be the set of all pizzas in the world right now. Find two sets, A and B, such that $A \cup B = P$ where A and B are disjoint.

951. Let P be the set of all pizzas in the world right now. Find two sets, C and D, such that $C \cup D = P$ where C and D are *not* disjoint.

Chapter Nine
Gardening Mall Announced

First part: Problems from this chapter

137. Dimensional analysis and unit analysis are fancy names for what we have been doing for a long time. It is called using a _____.
_{fill in two words}

(If you need a hint, it rhymes with *inversion tractor*.)
Wait a minute! What in the world is an "inversion tractor"?
Gulp. I'm not sure. I just made that up. Let me think. Okay. I know. It is an upside down tractor.

219. Lysander (played by Alex) has 476 lines to memorize. He could memorize 56 lines per hour. How long (in hours and minutes) will it take him to learn his part? (Use dimensional analysis.)

331. Gordon, who plays the front end of the donkey, has a speaking part.

    Jan is slightly jealous. She asks him, "How many lines do you get to speak?"
    Gordon frowned and said, "I only get to say two words. I say, 'Hee Haw.' "*
    For each performance Gordon gets paid $61.32 and Jan gets paid $44. Gordon gets paid what percent more than Jan? Round your answer to the nearest percent.

414. Are the set of words Gordon speaks in the play and the set of words that Jan speaks in the play disjoint?

---

* When you have a quote within a quote, the inside quote gets only single quote marks.

Chapter Nine   Gardening Mall Announced

Second part: the 𝔐ixed 𝔅ag: a variety of problems from this chapter and previous chapters

654. Find the LCD for $\frac{5}{6}$ and $\frac{2}{15}$

657. $\frac{5}{6} + \frac{2}{15}$

667. $\frac{5}{6} \times \frac{2}{15}$

    The director announced, "The show goes on tomorrow night. Our last dress rehearsal is tonight."

    Jan was puzzled. She asked Alex, "I thought that we would have weeks of rehearsal before opening night. How can we be performing tomorrow?"

    Alex smiled. "We have been rehearsing for weeks. We all know our lines."

    Jan was happy. She didn't have any lines to memorize. All she had to do was play the back end of the donkey. She was curious and asked, "Why did the director wait so long to cast my part?"

    Alex frowned. "You don't want to know."

    Jan insisted, "But I *do* want to know."

    "Okay. If you must know, Nell was originally playing the back end of the donkey. Yesterday was a dress rehearsal. Nell got sent to the hospital."

    Jan continued to insist. "Why? What happened?"

    Alex whispered, "Gordon farted."

    Jan got very nervous.

825. Part of the life in the theater is learning how to be creative. Jan traded 18 boxes of macaroni and cheese for a gas mask. If 3 boxes of m & c weighed 19 ounces, how much did all 18 boxes weigh? (Please use an ~~inversion tractor~~ a conversion factor.)

917. The director told Jan she would be paid $44 after tomorrow's performance. Jan needed the money right now. Edgar said he would lend her $35.20 right now and she could pay him $44 tomorrow. The interest charge would be $8.80. What percent interest is that?

Chapter Ten
At the New Mall

First part: Problems from this chapter

205. The dress rehearsal went from 6 p.m. to 10:25 p.m. How long was it?

252. The next morning Jan phoned you and said, "I've got a part in *A Midsummer Night's Dream*. We open tonight."
    You told her, "I'd love to see it. Are any tickets available?"
    Jan said, "They are normally $20, but as a cast member I can get a ticket for you at 20% off."
    How much will you be paying?

322. With the $35.20 she got from Edgar, Jan felt rich. For breakfast she headed out to her favorite *Desert-for-Breakfast* restaurant.

    She ordered four pieces of chocolate cake with raspberries.* The four pieces cost $30.25. The tax was 9%. What was the total bill?

463. At 9:20 she finished "breakfast." She headed home. At 10:07 she tossed her cookies.** How long is it from 9:20 to 10:07?

526. Jan slept from 10:07 a.m. to 5:45 p.m. She woke up and dashed to the theater. She knew that she didn't have to dress or comb her hair since no one would see her inside the donkey outfit. How long had she slept?

---

* The *p* in *raspberry* is silent.

** To toss your cookies means to keck. **Hold it! I, your reader, am lost. What does keck mean?** To keck means what some people call "making pizza." *I still don't get it. Are you doing that "Circular Definitions" thing from Chapter 3?* Yup. "Making pizza" means to regurgitate, and that means to throw up. *Now I get it.*

Chapter Ten   At the New Mall

Second part: the 𝕄ixed 𝔹ag: a variety of problems from this chapter and previous chapters

661. Gordon and Jan got into their donkey costume. Gordon promised that he wouldn't do you-know-what, so Jan didn't have to wear her gas mask.

Gordon climbed out of the costume and yelled at Jan, "You smell like . . . like . . . used chocolate cake with raspberries. I quit!"

Without having brushed her teeth and without having changed her clothes, which were stained with you-know-what, Jan didn't smell very nice.

She suddenly has a speaking role. She got a copy of the play and memorized her line. Since no one would climb into the costume with her, she just stood up and became a walking donkey.

You can guess what happened.

THE KITTEN Caboodle

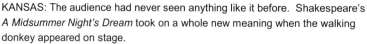

The Official Campus Newspaper of KITTENS University   Friday 11 p.m. Edition 10¢

Play Review

Donkey Steals the Show!

KANSAS: The audience had never seen anything like it before. Shakespeare's *A Midsummer Night's Dream* took on a whole new meaning when the walking donkey appeared on stage.

When the donkey appeared on stage left, the rest of the cast moved to stage right. Some of them were gasping. Besides delivering the line "Hee Haw," the donkey broke into a dance.

After the show, Jan was giving hoofprint autographs to all her fans. Four prints plus a $5.23 album to put them in were on sale for $16.35. How much was each hoofprint autograph worth?

Before you start the next problem, I, your reader, have a small question. I haven't spent years acting on stage. I'm not sure what "stage left" means. Who's left? When an actor is told to move to stage

Chapter Ten   At the New Mall

left, that means to the actor's left as he faces the audience. Of course, this is for English-speaking actors.

If the director, who stands or sits in the audience, is German-speaking, and tells you to cross* to stage left, it will be *from his point of view*, not yours. It's a different culture. If he tells you **cross stage left**, you had better move to your right.

If your director is French, and you are told to cross court side (côté cour), move stage left. Cross garden side (côté jardin) means move stage right.

Thanks. Now I know all the theater lingo.

You're joking, aren't you. Did you know that the stage you walk on is called the deck?

Do you know what crossing upstage means?

I give up. I'm not sure.

To cross upstage is to move to the part of the deck farthest from the audience. That's because in the old days the stage was raised so that actors in back could be seen more easily.

And crossing downstage is to head toward the audience.

The crusher position is also called full front. You face directly toward the audience.

$\frac{1}{4}$ left is to take a quarter turn to your left. When two people are talking on stage, they are often in $\frac{1}{4}$ left and $\frac{1}{4}$ right positions.

Left profile is rare since the audience has a harder time hearing you.

To upstage someone is to move upstage (toward the rear of the deck). That forces your fellow actor to assume the dreaded $\frac{3}{4}$ left position. That actor will hate you for doing that.

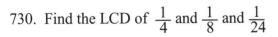

730. Find the LCD of $\frac{1}{4}$ and $\frac{1}{8}$ and $\frac{1}{24}$

---

* *Cross* is the technical term for *move* in stagecraft. A director will tell Jan to cross to stage left. That sounds more theatrical than just "move left."

Chapter Eleven
Seed Money

First part: Problems from this chapter

127. Ivy read the 11 p.m. edition of THE KITTEN CABOODLE newspaper. That gave her an idea for a new ice cream flavor: Midsummer Night's Cream.

One gallon of cream. Stir in a handful of chocolate cake and 3 raspberries. Fold in 5 ounces of whipped cream. Yum!

Cream comes in 10-liter containers. How many gallons is that? (Round to the nearest tenth of a gallon. One gallon is approximately 3.785 liters.)

236. 7.9 gallons of cream would require how many ounces of whipped cream?

409. Ivy's Ice Cream has always advertised: 1,000 flavors! She needed to eliminate one of her other flavors.

She had to decide between Buffalo Puffing and Sneaky Dog.

Ivy sold $\frac{4}{5}$ of a gallon of Buffalo Puffing this last month.

She has sold $\frac{7}{10}$ of a gallon Sneaky Dog this last month.

Which flavor has sold less?

501. Some of her customers renamed Buffalo Puffing as Puffy Buffy. Ivy liked that and adopted the new name. Puffy Buffy is spicy. A serving has 39% more cayenne than a one-pound burrito, which has 1.3 grams. How much cayenne does a serving of Puffy Buffy have?

Chapter Eleven   Seed Money

Second part: the 𝔐ixed 𝔅ag: a variety of problems from this chapter and previous chapters

656. One of Ivy's most beautiful ice creams was a golden yellow. It was a delight to the eye. *Zea mays* sold $2\frac{1}{4}$ gallons last month. How much more was that than the $\frac{7}{10}$ of a gallon that Sneaky Dog sold?

732. Ivy put 18 (identical) cobs of *Zea mays* into a 60 gram bucket. The whole thing weighed 834 grams. How much did each cob weigh?

900. One of the most famous formulas from algebra is d = rt (distance equals rate times time). Ivy's store is 52 feet wide. Ivy bought a special robot to carry Puffy Buffy ice cream, because it was so dangerous. It could run from the left side of the store to the right side at 8 feet per second. How long would that take?

918.  A man walked into Ivy's Ice Cream and ordered a triple scoop of Puffy Buffy. Ivy had purchased a special warning sign ➻ but the man ignored it.

He ate the whole thing in three seconds. He melted.

He was 162 pounds. He was now 18 pounds, which was the weight of his ashes. What fraction of his original weight is he now?

DANGER
CORROSIVE MATERIALS,
WEAR REQUIRED
PROTECTION

Puffy Buffy should be eaten very carefully!

Chapter Twelve
Getting Things Arranged

First part: Problems from this chapter

152. After her first performance in *A Midsummer Night's Dream*, Jan had received the biggest applause of all the cast members. She stood downstage in crusher position and bowed and bowed.

   She headed back to her apartment, took a shower, brushed her teeth, and fell asleep with a smile on her face. She dreamed that the director would let her play the lead role of Hermia or Helena.

   You phoned her at 10 a.m. the next morning and woke her up. Jan had gone to sleep at 11:17 p.m. How long had she slept?

325. You agree to meet for lunch. Jan suggested heading to Stanthony's PieOne at 1 p.m.* "I'll pay for lunch this time," Jan announced. She had sold 100 of her $16.35 four hoofprints-plus-album packages last night. How much had she received? (Don't use a calculator on this problem!)

460. At Stanthony's you noticed that he was offering a new pizza: Theatrical Arts Pizza. It was pure ham.

   The 16-inch pizza was $5.76. (A 16-inch pizza means that its diameter is 16 inches.) What was the area of this pizza? (Use 3 for π in this problem. $Area_{circle} = \pi r^2$ where r is the radius.)

541. In percentage terms how much bigger in area is Stanthony's 20-inch pizza than his 16-inch pizza? (Again, use 3 for π. The answer is not 25%.)

---

* What hours you keep depends in part on what job you have.

☞ Playground directors and people who sing and dance at ice cream stores typically work during the daylight hours.

☞ Theatrical arts performers tend to work later in the day because performances are usually in the afternoon or in the evening.

☞ I, your author, do my best writing from 4 a.m. to 6 a.m. each day. It's generally pretty quiet during those hours. Some other authors prefer to work late at night.

Chapter Twelve   Getting Things Arranged

Second part: the 𝔐ixed 𝔅ag: a variety of problems from this chapter and previous chapters

652. After lunch you and Jan decided to head to Ivy's for old time's sake.

With 1,000 flavors, Ivy's menu was the size of a book. If she listed 23 flavors on each page, how many pages would her menu be?

773. Scientists who study human behavior have learned that if you offer people too many choices, they tend to go a little nuts.

Jan just flipped the menu open at random and pointed to the first item on that page. "I'll have a milkshake with that flavor," she said.

A minute later a special waiter delivered her shake. Jan gulped it down while she was talking to you, "I really enjoyed singing and dancing in *A Midsummer Night's Dream.* They gave the whole stage to me. I felt very special." Suddenly her voice disappeared. (Puffy Buffy can have that effect.)

Jan couldn't even say, "Hee Haw."

She wrote on a piece of paper, "I can't talk. I have to perform tonight. Get me to the throat hospital. Now!"

You had never heard of a "throat hospital" before. Obviously, Jan was panicking.*

It was 3.2 miles to the nearest hospital. You drove Jan at 10.2 mph (miles per hour). How long would it take to get there? Give your answer to the nearest minute.

836. You couldn't find the emergency entrance so you parked at the front door.
Jan said that she wasn't a veteran, but you assured her that in an emergency they would see her.

---

* English spelling can be a little weird. *To panic* means to have a sudden giant fear. *Panicking* had a *k* in it.
  picnic ➡ *picnicking*
  frolic ➡ *frolicking*
  traffic ➡ *trafficking*

33

Chapter Twelve   Getting Things Arranged

You headed into the waiting room and told Beth, "She's got a bad throat problem. She's going to be a donkey tonight and she can't say 'Hee Haw.'"

Beth smiled and said, "We'll get her hee-hawing in no time. Please take a seat in the waiting room."

Sitting in the waiting room, Jan noticed that this didn't seem like a typical veterans' hospital.*

After 45 minutes you and Jan were called into the examination room. Beth looked at you and at Jan and asked, "Where's the donkey?"

Jan raised her hand.

Beth shook her head and said, "The set of all patients we have here at this veterinary hospital and the set of all human beings are _____."
_{fill in one word}

902. You explained to Beth that Jan is the lead actor (stretching the truth a little bit) in *A Midsummer Night's Dream*, her throat is really sore, and she can't speak.

"Poor dear" Beth said to Jan and patted her on her head. (Beth is used to treating cows, not people.) "This will cool your throat," she said as she offered Jan a sip of her milkshake. It was Ivy's Midsummer Night's Cream milkshake—the least irritating milkshake that there is.

It worked.

Jan was overjoyed. She gave Beth a hug and danced out of the hospital. Theater arts people are often more flamboyant than mathematicians.

When Jan and you got back to your car, she said, "Well, what are we going to do this afternoon?"

Before you could answer, Jan said, "Let's go bowling."

Let set A = the set of all possible things to do on an afternoon. Let set E = the set of activities involving eating. Let B = the activity of bowling. Does A = E ∪ B?

---

* Did you notice where the apostrophe went? If there were only one veteran, it would be veteran's hospital. If you have six dancers and you want to talk about their costumes, it is dancers' costumes.

Chapter Thirteen
Blood and Your Brain

First part: Problems from this chapter

130. You explained to Jan, "It's been years since I've gone bowling."
She grinned and said, "I've never been bowling. It will be a new adventure for me. I crave adventure."

You thought to yourself, *I wonder if she'll live to the age of 30 at the rate she's going. At least bowling is fairly safe. I'm glad she doesn't want to spend the afternoon hunting bears.*

You arrive. Your rent the bowling shoes. You show Jan where the bowling balls are.

Jan tells you, "You go first. I want to see how it's done."

 You start with 10 pins. You knock over one pin. What percent of the pins remain standing?

241. "My turn!" Jan yells.
You explain to her that in bowling you get two tries and then it will be her turn. On your second try you knock down the other nine pins. What percent of the pins remain standing?

431. Jan takes her turn. She knocks down six pins. "Okay, I get it. So this is bowling?" She starts to head back to the shoe rental place.
You ask, "What are you doing?"
Jan frowned. "What do you mean? Aren't we done? You had a turn, and I had a turn. Is there anything more to bowling?"
"But we are each supposed to have ten turns!"
Jan had a simple question, "Why?"
You didn't have a really good answer. You follow her. Both of you turn in your shoes. $12 for 4 minutes of bowling. How much is that per hour? (Use a conversion factor.)

Chapter Thirteen   Blood and Your Brain

Second part: the 𝔐ixed 𝔅ag: a variety of problems from this chapter and previous chapters

557. You want to explain to Jan that some people spend a whole afternoon bowling, and they do it every week. They laugh and shout when they knock down all the pins. They write down on paper how many pins they have knocked down.

But you are afraid of what Jan might ask. "Why?"

If you say, "Bowling is relaxing," she'll respond with, "Napping is better."

If you say, "You can get a trophy if you knock over a lot of pins," she'll respond with, "And what good is a stupid trophy?"

If you say, "Bowling gives you a break from your normal routine," she'll say, "And going bowling every Thursday afternoon for twenty years with your buddies...." She wouldn't even have to finish that sentence.

Jan's eyes light up. "Let's do something that can give some real results. Let's go bear hunting. Three hundred pounds of fresh bear meat will feed us for a half year."

Do Jan's numbers make sense—or is she exaggerating?

653. "Have you ever been bear hunting?" you ask. This is an 𝕚𝕞𝕡𝕠𝕣𝕥𝕒𝕟𝕥 𝕢𝕦𝕖𝕤𝕥𝕚𝕠𝕟 for you.

"No," Jan answers. "But it will be a new adventure for us. I crave adventure."

You have heard this from Jan before. A thousand thoughts cross your mind:

1. I crave living! I want to live a long and healthy life.
2. Jan managed to get injured in an ice cream store.
3. Jan doesn't know anything about bear hunting.
4. Faster humans can run about 16 mph for a hundred yards. A bear can run about 25 mph over a distance of two miles.

What percent faster is a bear than a human? (Round your answer to the nearest percent.)

734.   5. Bears can eat all kinds of meat. They will eat moose, deer, sheep, bison, and theater arts majors. They will even eat friends of theater arts majors.

You say, "Jan, let's go shopping instead. That can be a lot of fun."

You wait. Jan thinks. She says, "Okay." You are relieved.

You drive $\frac{3}{8}$ of a mile to the New Mall. You park and walk $1\frac{3}{4}$ miles to the nearest store. (big parking lot) How far have you traveled?

Chapter Fourteen
Conversion Factors

First part: Problems from this chapter

153. The nearest store is Abby's Agricultural Supplies. You and Jan see this little square-headed kid running around in the store. He's holding a little shovel and bag full of money.
    Jan thought that looked a little weird. You head on to the next store.
    Harry's Handy Hardware looked inviting to you, but Jan didn't like the thought of hairy hands.
    Neither you nor Jan wanted to go into Ivy's Ice Cream, Store #2.
    Julie's Jewelry seemed inviting to both of you. As you head inside Jan says, "I need a ring."

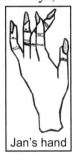
Jan's hand

You look at her hand and ask, "Where could you wear another ring?" Jan knew the answer. "I have only one ring on my thumb."*
    Julie showed Jan "the perfect ring." She hoped that Jan wouldn't notice the big T on the ring. The ring had been made for Thadeus, but he had changed his mind and never paid for it.
    "It's a Theatrical arts ring! I'll take it." Jan didn't ask how much, so you did. Julie said it would "only be $34 per month for 1,044 months."
    How many years is that?

224. How much would the total cost be?

---

* The *b* in *thumb* is silent . . . like the *b* in *dumb*. Wouldn't it be reasonable to call this 🥁 a drumb?

Chapter Fourteen   Conversion Factors

Second part: the 𝔐ixed 𝔅ag: a variety of problems from this chapter and previous chapters

370. "My wrist hurts," Jan complained. That is no surprise. She has 8 pounds of rings on her hand. If she is wearing 13 rings, how much does each ring weigh? (Assume all the rings weigh the same. Round your answer to the nearest ounce. 1 pound = 16 ounces.)

411. It was time to leave Julie's Jewelry. One jewelry purchase of $35,496 was enough—even though Jan kept saying that it is was only $34 per month.

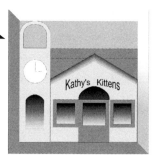

Jan screamed, "This is the store!"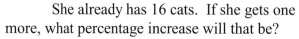
Theater arts people scream more often than mathematicians.* "I *need* a cat!" She may have even said, "Hee Haw," but that's not certain. In any event, Jan was excited.

She already has 16 cats. If she gets one more, what percentage increase will that be?

564. Kathy's Kittens was the perfect store for people who really like cats. In order to make shopping quick and easy, Kathy offered several options:

Set A = {a gray cat, a quart of Ralph's litter, a scoop}
Set B = {a black cat, a quart of Ralph's litter, one leash}
Set C = {a gray cat, a cat toy, a collar}
Set D = {a black cat, a scoop, a cat toy}

Which two sets are disjoint?

---

* But not always. The king asked Archimedes how to determine whether his crown was pure gold. When Archimedes was taking a bath, he suddenly figured out how to figure out how to do that without cutting the crown. In Greek he shouted, "Eureka!—I found it" and ran through the town naked. Figuring out some math problems can be really exciting.

Chapter Fifteen
Plants Don't Eat Dirt

First part: Problems from this chapter

106. Jan bought set A = {a gray cat, a quart of Ralph's litter, a scoop}. With all her cats she could always use another quart of Ralph's litter.

A quart of litter comes in a 3-pound box. There was a hole in the box and 5 ounces had leaked out. How much was left?

250. Five ounces is what percent of 3 pounds? Give your answer to the nearest percent. (The answer is not 60% or 167%.)

371. You carried the litter and the scoop. Jan carried her new gray cat, and you both headed to a mall bench to sit down. Jan played with her cat (and you were allowed to play with the scoop if you wanted to).

She tried to pet her cat, but all the rings on her hand scared the cat. It scratched Jan's face and ran away.

Jan wrote up a reward poster.

If someone turned in ten gray cats, how much reward might they get?

(The answer is not fifty dollars or five dollars.)

455. Before getting this gray cat Jan had 35 scratches on her face and arms. She now had 20% more. How many scratches does she now have?

542. It was time to end the shopping trip. You wanted to head back to Jan's apartment and see her 16 cats. You told Jan, "I didn't know that someone could have that many cats."

Jan retorted, "But Kathy (of Kathy's Kittens) has more than I do."

"But she sells them" you explain. "If you sold your cats for $4 each, think of all the money you could make." (Do this problem in your head without writing anything down.)

Chapter Fifteen   Plants Don't Eat Dirt

Second part: the 𝔐ixed 𝔅ag: a variety of problems from this chapter and previous chapters

592. When you got to Jan's apartment, you counted her cats. There were 21 of them, some big and some very small.
"Jan, you told me you had 16 cats. Did you miscount?"
She hadn't miscounted. She had even named all her cats: Able, Beatrice, Calloway, Dunkirk, Edna, Francis, Galloway, Hinkenmeyer, Icabod, Juno, Kilpatrick, Lumpy, Mouse-face, Needles, Oops, and Padunk.
Where did the other five come from?

725. $(4\frac{3}{7})^3 = ?$

863. For what values of x will 1^x equal one?

949. All 21 cats rushed up to you and purred. You were still carrying the quart of Ralph's litter. Jan hadn't changed the litter boxes in a week. You opened a window.
You poured the 2 pounds, 11 ounces of Ralph's litter into the three litter boxes, dividing it evenly.
Seven cats lined up at each box awaiting their turns.
How many ounces of litter did each box receive?

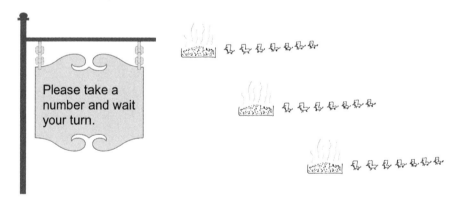

Chapter Sixteen
Out of Thin Air

First part: Problems from this chapter

132. Cats eat meat. They are carnivores (meat eaters). In fact, they are obligate carnivores—they must eat meat.*

They enjoy a tasty mouse. The carbon in the mouse-meal combines with the oxygen that the cat has breathed in. That creates the energy for the cat. It also creates carbon dioxide (CO_2), which the cat breathes out.

Plants love CO_2. Without it they die. Plants, like wheat, "breathe" in CO_2 and grow. Mice eat the wheat. Cats eat the mice. That carbon keeps making a round trip: cat → wheat → mouse → cat → wheat → mouse → cat → wheat → mouse → cat → wheat → mouse → cat → wheat → mouse → cat → wheat → mouse → cat → wheat → mouse → cat → wheat → mouse → cat. . . .

One out of every 2,000 air molecules is CO_2. That's 0.05%. If the carbon dioxide concentration were 0.1%, then plants could find a CO_2 in one out of every how many air molecules?

255. Let set A = the set of Jan's cats that weigh more than 5 pounds.
Let set B = the set of Jan's cats that weigh less than 8 pounds.
Let set C = the set of Jan's cats that weight more than 6 pounds.
Question 1: Does A ∪ B equal the set of all of Jan's cats?
Question 2: Is A a subset of B?
Question 3: Is C a subset of A?
Question 4: If set D is a subset of set E and set E is a subset of set F, must it always be true that D is a subset of F?

---

* Dogs eat both meat and plants. One dog that I know enjoys ice cream, although it isn't very good for him because of the sugar.

The bodies of dogs are different than the bodies of cats. Their tummies and intestines are different. Officially, their digestive systems are different.

Cows and horses eat only plants—herbivores.
Cats eat only meat—carnivores.
Dogs and people can eat both—omnivores.

Chapter Sixteen   Out of Thin Air

Second part: the 𝔐ixed 𝔅ag: a variety of problems from this chapter and previous chapters

323. Charlie is the name of a cat that lives next door to Jan.
Let C = {Charlie}
Let J = {Able, Beatrice, Calloway, Dunkirk, Edna, Francis, Galloway, Hinkenmeyer, Icabod, Juno, Kilpatrick, Lumpy, Mouse-face, Needles, Oops, Padunk, Quincy, Rats, Slender, Tagalong, and Uranus}, which is the set of all 21 of Jan's cats.

The way to prove that C is not a subset of J *is to find a member of C that is not a member of J.* That's the easy way to show C is not a subset of J.

Must the empty set, { }, be a subset of J?

432. If set M is a subset of set N, then what must M ∪ N equal?
(Some of my questions are easy, and some are tough.)

627. Is the set of herbivores a subset of the set of omnivores?

735. Jan has pink hair 5.37% of the time. What percent of the time does she not have pink hair?

865. One of her cats, Juno, was chewing on your ankle. She was hungry. Jan had forgotten to feed her cats. The other 20 cats had leaped out the window that you had opened when you first came in Jan's apartment. They went looking for food.
    What percent of Jan's cats remained in her apartment? (Round your answer to the nearest tenth of one percent.)

Chapter Seventeen
Photosynthesis

First part: Problems from this chapter

151. Looking at wheat → mouse → cat → wheat → . . . we can think of it as a giant circle of life. The plants combine carbon dioxide and water and produce sugars, starches, and oils. Animals eat the sugars, starches, and oils and give off carbon dioxide.

    Like a wheel that it spinning, it would come to a stop unless something keeps it spinning. Where is energy inserted into this wheel of life?

273. After Juno finished chewing on your ankle, she lay down and took a nap.* Charlie, the cat that lives next door, came through the window. He had heard that there was good ankle chewing going on and didn't want to miss any of the action.

---

* She *lay* down. The past tense of *lie* is *lay*.
    Yesterday I lay in bed for three hours.
    Yesterday I laid my backpack down on my bed.
        The past tense of *lay* is *laid*.
English ain't easy. I think it's harder than math.

Chapter Seventeen Photosynthesis

Charlie walked up to you and asked, "Do you mind if I take a little bite?"

You screamed.

Three of Charlie's bites plus 0.7 pounds of his spit weigh a total of 6.1 pounds. How much is each of Charlie's bites?*

Charlie

324. "Oh! Lookie!" Jan exclaimed. "I have a new kitty."

Jan had one cat (Juno). She now has two cats. What percent increase did she experience?

433. Charlie is an obligate carnivore. He won't eat your breakfast cereal. He won't eat your potato chips. However, he will eat ■■■. (This has been censored to keep this book G-rated. I don't want you or Jan to have nightmares. Both y-o-u and J-a-n have three letters.)

Only living things can do photosynthesis. Can Charlie do photosynthesis?

---

* How much **is** each of Charlie's bites?
  The subject of the sentence is *each*.
  Each is.

  of Charlie's bites is called a prepositional phrase.

Examples of prepositional phrases:
  into the forest
  out of my mind
  around the corner
  beside the still waters
  with extra anchovies

Chapter Seventeen   Photosynthesis

Second part: the 𝔐ixed 𝔅ag: a variety of problems from this chapter and previous chapters

565. $(3\frac{1}{3})^3 = ?$

629. What is the cardinal number associated with {Jan, Juno, Charlie}?

733. If set A = {red, yellow, green} and set B = {orange, blue}, then does card A + card B = card (A ∪ B)?
      (**card A** means the cardinal number associated with set A.)

To do these next two problems, you are going to have to be a little creative. You will have to invent sets C, D, E, and F. I'm not going to tell you what will go in each set.
    If you like baseball, you could put famous players in those sets.
    You could use pizza toppings.
    You could use the names of your eight children: Alice, Booboo, Candy, Didi, Eileen, Fusty, Grindle, and Happy.
    You probably won't be able to tell the answers just by looking. Take out a piece of scratch paper and play.

837. If sets C and D are disjoint, will it always be true that
      card C + card D = card (C ∪ D)?

930. If sets E and F are *not* disjoint, which of these will be true:
      i) card E + card F < card (E ∪ F)   < means "less than"
      ii) card E + card F = card (E ∪ F)
      iii) card E + card F > card (E ∪ F)   > means "greater than"

Chapter Eighteen
Fred the Heterotroph

First part: Problems from this chapter

157. You take a slurp of Ivy's SUNSHINE IN YOUR TUMMY ice cream. It starts in your mouth.
  Arrange the rest of these in the order that the ice cream visits them:
    small intestine
    pharynx
    tummy
    large intestine
    esophagus

274. Are you an autotroph?

329. Which one of these is the longest: mouth, pharynx, esophagus, tummy, small intestine, or large intestine?

461. The stomach, pancreas, liver, and gallbladder dump about 9 liters of fluid into the first part of the small intestine each day. (1 liter = 1.057 quarts). How many quarts is that? Give your answer to the nearest tenth of a quart.

591. The small intestine was the Place of Final Digestion or the Welcome Center. Given the fact that 9 liters of fluid is being dumped into that twenty-foot tube every day to mix with the contents of your stomach, I would also call the small intestine the **WETLANDS**. That soupy mixture makes it really easy for the good stuff (sugars and amino acids) to move from the small intestine into your blood stream.
  Everyone knows that 9 liters equals $\frac{9 \text{ liters}}{1} \times \frac{1 \text{ gallon}}{3.785 \text{ liters}}$ ≈ 2.4 gallons. I bet you don't drink 2.4 gallons every day. Why don't you have to do that?

Chapter Eighteen   Fred the Heterotroph

Second part: the 𝓜ixed 𝓑ag: a variety of problems from this chapter and previous chapters

640. Jan looked at the clock. It was 4:40. She had to be at the theater by 6:10. How much time did she have?

755. Charlie saw a rabbit out on the lawn in front of Jan's apartment. He leaped out the window and chased the rabbit. You quickly shut the window. Jan had had two cats.* She now had one. What percent change had she experienced?

> ### Intermission
>
> Long ago when I was in middle school, I wrote an essay. In it I wrote that Henry had had years of pain before he had gone to the doctor.
> Another student, whom I will call Thud, copied my essay, but wrote, "Henry had years of pain before he had gone to the doctor." Thud didn't know about the past perfect.
> Without punctuation I could truly say that Stan while Thud had had had had had had had had the teacher's approval.
> With punctuation: Stan, while Thud had had "had," had had "had had" had the teacher's approval.

827. For many students, the hardest part of algebra is word problems, in which English gets stirred in with math. Plain algebra is much easier. Solve $6x + 43 = 145$.

---

* Jan had had two cats. In English there are not just three tenses: past, present, and future. When Charlie left, Jan had one cat. That's in the past tense. To talk about something that happened before "Jan had one cat," we use the past perfect tense. Jan had had two cats (past perfect) before she lost (past tense) one of them.

In *Life of Fred: Classes*, which is the third book in the high-school English series, we will give lots of examples and explain all of the twelve (12!) tenses.

Chapter Nineteen
Eyes, etc.

First part: Problems from this chapter

122. It was 4:40. Jan put on her face powder, her lipstick, her rouge, her eyeliner, her fake eyelashes, her nail polish, her toenail polish. She dyed her hair and put on her pearl necklace.

You asked, "Why? You are going to be in a donkey costume."

Jan had the standard theater arts answer, "Suppose Cassie, who plays the part of Hermia, gets sick. I know that role. I could play Hermia and be the star of the show."

You say, "I know Cassie. She's in perfect health. She doesn't drink or smoke. When she drives, she wears her seatbelt and has the radio off so that she doesn't get distracted. When she plays cards, she likes to use an older deck so there is less chance that she will be cut by a sharp-edge card."

"Well, we actors have to be ready for our break when it comes. There's a 0.2% chance Cassie won't be able to perform."

That is one performance out of how many?

> **Another Intermission**
>
> You asked Jan, "But if you get to play Hermia, then who will play the donkey?
> She looked at you but didn't say a word.

222. We are all in the *Dance of Life*. Some of us, like Jan, like to sing and dance around on a stage and hear the applause. Ten minutes of hearing people clap, and then they take off their stage makeup and go home.

Others have many other parts in that *Dance*. Some collect garbage (sanitation workers). The world would be a very smelly place without them. Some are parents of large families. Some watch television.

When Jan looked at you and didn't say a word, you shook your head. Being an ass on the stage wasn't what you are called to do in the *Dance of Life*. Jan dashed out the door and left

Chapter Nineteen   Eyes, etc.

you to turn off the lights. You turned off the living room light, the bathroom light, and the kitchen light. Suddenly, **it was dark.** Match each action in the first column with the receptor in the second column.

1. You heard the refrigerator.
2. You smelled the kitty liter boxes.
3. You knew that you were standing straight up.
4. You felt a fly land on your forehead.
5. You couldn't see a thing.

A. mechanoreceptor
B. proprioceptor
C. photoreceptor
D. audioreceptor
E. chemoreceptor

327. As you felt your way toward Jan's front door, you tried to remember where all her furniture was. You accidentally stepped on Juno, who had been sleeping after her snack on your ankle.
    Which of your receptors could tell you what you had done?

466. You got to the door and left her apartment. Charlie was there with half of a rabbit in his teeth. You didn't know whether to feel happy for Charlie or sad for the rabbit.
    Charlie didn't say anything, because his mother had told him never to talk with his mouth full.*
    You walked the 2,000 feet to the theater. Actually, you limped because of your ankle. You walked at the rate of 4.5 feet per second. How many seconds did it take you? (Round your answer to the nearest second.)

562. Convert your answer to the previous problem into minutes and seconds.

---

* Do you think I'm kidding? Tell me the last time you ever saw a tiger talking with his mouth full. Tigers listen to their mothers.

Chapter Nineteen   Eyes, etc.

Second part: the 𝔐ixed 𝔅ag: a variety of problems from this chapter and previous chapters

593. When you got to the theater, you realized that you were much too early. The play was scheduled to begin at 7 p.m. The mall had gone dark for some reason. You wondered where to go to spend some hours before the show.

The Kansas public library! You remembered the sign that was posted on their front door:

> 1. No gum chewing.
> 2. No gum in your mouth even if you aren't chewing.
> 3. No playing tag.
> 4. No leaving gum inside the books.
> 5. No tigers allowed.

That rule #5 was the one that everyone liked (except Charlie). Libraries are one of the safest places on the planet.

libraries are happy places

You had spent many happy hours running up and down the aisles (remembering not to play tag). You had found many interesting books to take home to read. And it was free.

What was weird were the muddy footprints that were headed right toward you.

You wondered why it was only the right foot. You noticed that it wasn't mud; it was blood. You didn't feel good. Things were spinning.

The library looked strange.

You added up one plus one and got two.
  1. ≫ Juno had attacked your right ankle twice.
+ 1. ≫ You had lost lots of blood.
  2. ≫ You passed out.
Humans normally have about 5 quarts of blood. You have lost 1.7 quarts. What class of hemorrhaging (bleeding) are you experiencing? (See next page.)

Chapter Nineteen   Eyes, etc.

This is the actual chart from the American College of Surgeons' Advanced Trauma Life Support group. You'll notice they have their apostrophe in the right place.

Class 1: Less than 15% of the body's blood lost.
When you donate blood, about 9% of your body's blood is removed. Usually there are no symptoms with less than a 15% blood loss.

Class 2: 15 < blood loss < 30%
A bit weak. Pale skin. Skin a little cold. Heart rate goes up.

Class 3: 30 < blood loss < 40%
In shock. Heart rate is fast. Blood pressure drops.

Class 4: A loss of > 40%
You are about to find out everything you ever wanted to know about God and Heaven.

666. In the ambulance they were doing the medical stuff so that you didn't enter Class 4. Among other things, they stopped your bleeding. They stuck on bandages. They charged you $5.24 for each bandage. It cost them a nickel (5¢) for each bandage. They applied 19 bandages. Since there wasn't enough room on your ankle for all those bandages, they put some on your tummy. Your tummy didn't need any bandages, but they put them there "just in case."
   How much were you billed for those bandages?

756. If you start with 5 quarts of blood, how much would you have to lose to have lost 40%? Do this problem in your head and just write down the answer. No calculating with a pencil. No need for a calculator.

866. You had $2,000 in your checking account. The final bill was for 80% of the $2,000. How much was the bill? Again, no need for a calculator. Just write down the answer.

Chapter Twenty
Negative Numbers

First part: Problems from this chapter

133. Cassie had carefully walked the two blocks from her house to the theater. She looked all four ways before she crossed the street. In fact, she looked all four ways twice. "You can never be too careful," she thought to herself.

She didn't walk on any of the cracks in the sidewalk, because she didn't want to accidentally twist her ankle. She knew that her playing the role of Hermia was critical. No carelessness on her part would stop her.

She didn't answer the telephone after 4 p.m. (the performance was at 7) so that she could save her voice. And, of course, a large scarf to protect her throat.

And Charlie got her. Her last thought was, "Who in the world could have planned on meeting a tiger on the streets of this town?"

Charlie was the 0.2%.

Jan got the part of Hermia, which she had dreamed about. Cassie's aunt, who had come to see her dear niece perform, was drafted to play the role of donkey. It took her four minutes to memorize her line, "Hee Haw."

They let you out of the hospital. You arrived half way through the second act (of five acts). What percent of the play had you missed?

221. Jan saw you come in and sit down. She waved to you. You are not supposed to do that when you are acting on stage. Her salary for the night was $87. They fined her $100 for her waving. Her gain for the evening was a negative number. What was that number?

349. –73 is a small number. It is smaller than zero. You would rather have $0 in your checking account than have –$73. You would have to deposit $73 in that account in order to get it up to a balance of zero.

$-73 < 0$

Name a number that is smaller than –73.

52 × 3 = 156   52 × 4 = ...

52) 1950

52 wks of a year

38 wks of a year

38/52

1. esophagus, pharynx, tummy, small intestine, large intestine.

2. No, I am a heterotroph

3. Small intestine

4.
```
    5 6
  1.0 5 7
  ×   9
  ─────
  9.5 1 3
```
[9.5 quarts]

5. Because just like the heart makes blood, we make fluids as well.

[1:30 minutes]

6.
```
   5 18
   6̸8̸ 0
  - 4 4 0
  ───────
     1 6 0
```

7. 50%

8) $1x + 43 = 145$

$\dfrac{6}{6} \times 1x$

$y \neq \dfrac{7}{6} = 145$

$6\overline{)43} \quad 7\frac{1}{6}$

6

1. $\frac{2}{5} = \frac{40}{100} = 40\%$

2. $\frac{9 \times 10}{9 \times ?} \quad \frac{27}{?} \quad | \quad -0|3$

3. -74

4. $\frac{9 \times 8}{9 \times 10} \quad \frac{72}{?} \quad | \quad -13$

5. $\frac{120}{500} - \frac{?}{?} \quad | \quad -19$

Chapter Twenty Negative Numbers

Second part: the 𝔐ixed 𝔅ag: a variety of problems from this chapter and previous chapters

457. Arriving in the middle of the second act, you missed Hermia's (Jan's) entrance in the first act. Jan was so excited about her getting to play one of the main roles that she said, "Hee Haw." The director fainted.
    He fined her $120 for forgetting her lines.
    So far she has a salary and two fines: $87 – $100 – $120.
    That's the same as 87 – 220. What does that equal?

563. In the third act the donkey comes on stage. Tonight Cassie's aunt had hoped to see her niece in one of the main roles. Instead, she was having to play the donkey. She kept saying "Hee Haw." She had memorized her line, but she wasn't sure when to say it.
    Also in the third act, Hermia (Jan) was supposed to have an argument with her girl friend Helena. Jan got so excited that she hit Helena. This was not good. Helena did not hit her back. Helena did not get a $500 fine.
    Jan had made an extra $12 for teaching Cassie's aunt her lines. She had also earned three fines.
    The best way to combine together +87 – 100 + 12 – 120 – 500 is to combine together the positive numbers. Then combine together the negative numbers. Then combine the one positive number with the one negative number. Do it.

628. After the play Cassie's aunt was exhausted but smiling. She had never acted before. She wondered why Cassie had not shown up this evening. Everyone knew that Cassie led a very careful life.*
    Cassie's aunt would never learn what happened to her niece for 1,650 weeks—until her granddaughter read Chapter 20 of this *Zillions of Practice Problems* book and told her grandmother. How many years and weeks is 1,650 weeks? (Assume a year contains 52 weeks.)

---

* You can be too careful, and you can be too adventurous. The middle road is often the best.
    On Monday, Wednesday, and Friday my mother would tell me, "Look before you leap." On Tuesday, Thursday, and Saturday she would tell me, "He who hesitates is lost."

Chapter Twenty-one
Eyelashes

First part: Problems from this chapter

154. Cassie had short eyelashes. Her parents both had long eyelashes. What is the probability that Cassie's brother had short eyelashes?

253. Cassie had short eyelashes. Must one of her great-great grandparents have had short eyelashes?

328. Cassie had short eyelashes. Must one of her great-great grandparents have had an S gene?

413. Cassie had short eyelashes. Must at least two of her great-great grandparents have had an S gene?

575. You (not Cassie) have long eyelashes. You have 16 great-great grandparents. Is it possible that only one of those 16 had long eyelashes?

668. There was a time in which the earth didn't exist. Everyone agrees upon that fact.

When you look at your ancestors back in history,
your parents
    your grandparents
        your great grandparents
           your great-great grandparents
              your great-great-great grandparents
                etc.
there had to be a point at which this began. (Otherwise, you would have ancestors but no earth for them to stand on.)

There couldn't have been just one super-earliest ancestor, because then there wouldn't have been any kids.

Two is a nice number to start the Human Family. What do we know about their eyelashes?

Chapter Twenty-one   Eyelashes

Second part: the 𝓜ixed 𝓑ag: a variety of problems from this chapter and previous chapters

760. Name the four subsets of {Cassie, Charlie}.

824. Let's think for a moment about the subsets of {Cassie, Charlie, Jan}. We know that there are 4 subsets of {Cassie, Charlie}. We just did that in the previous problem.
    Take each of those 4 subsets and add in Jan. For example, {Charlie} would become {Charlie, Jan}. That creates 4 more subsets of {Cassie, Charlie, Jan}. So the total number of subsets of {Cassie, Charlie, Jan} is 8. (Hint: 4 + 4 = 8)
    How many subsets does {Cassie, Charlie, Jan, Ivy} have?

828. How many subsets does {A, B, C, D, E, F} have?

862. $(4\frac{1}{6})^2 = ?$

919. Some people with short eyelashes think that they look more attractive if they have long eyelashes. One solution is to have had different parents, but that is fairly hard to do.
    The other solution is to buy fake eyelashes and stick them on your eyes every morning.

before   after

    You can buy a bucket of Ralph's eyelashes. It's a can (a cylinder) that is 4 inches tall with the radius of 2 inches. What is its volume? $V_{cylinder} = \pi r^2 h$ Use 3.1 for π and round your answer to the nearest cubic inch.

970. Using your answer to the previous problem and the fact that one Ralph eyelash has a volume of $\frac{5}{134}$ cubic inches, how many of them are in a can?

Chapter Twenty-two
Variation

First part: Problems from this chapter

156. What number(s) make 0x = 0 true?

257. What number(s) make 0x = x true?

373. What two numbers make $5x = x^2$ true?

434. Solve 3x + 1.9 = 10. (Hint: You can't find the answer just by guessing numbers. Use the procedure you used in solving #827 or #855 or #206 in the Complete Solutions and Answers section of this book.)

580. Six buckets of Ralph's eyelashes plus a $2.98 tube of eye glue cost $248.98. How much does a bucket of Ralph's eyelashes cost?

    We solved the first three equations on this page by guessing numbers.
    We solved the last two equations on this page by using algebra. (We subtracted a number from both sides. We divided both sides by a number.) These last two equations would be hard to solve by guessing, but were easy to solve by algebra.
    That's why algebra was invented.

Chapter Twenty-two   Variation

Second part: the 𝔐ixed 𝔅ag: a variety of problems from this chapter and previous chapters

645. Many people couldn't afford a bucket of Ralph's eyelashes because it cost $41. He introduced his new cone-of-lashes.

The cone had a radius of 3 inches and a height of 4 inches. To the nearest cubic inch, what was its volume?

Use 3.1 for π. $V_{cone} = (\frac{1}{3})\pi r^2 h$

800. In the previous chapter we learned that a bucket (cylinder) of Ralph's eyelashes was about 50 cubic inches. It cost $41.

In the previous problem we learned that a cone-of-lashes had a volume of 37 cubic inches. Using a conversion factor, find out how much Ralph should charge for his cone-of-lashes.

835. From the previous problem you know how much Ralph should charge for his cone-of-lashes. He decided to sell it with a 19% discount. To the nearest cent, how much would he charge?

864. Since everyone was wearing fake eyelashes, Jan decided to be different. She put a fake eyelash on her pinna. Instead of calling it an eyelash, she called it an _____.
fill in one silly word

977. Why can't you put a fake eyelash on your retina?

57

Chapter Twenty-three
Blood

First part: Problems from this chapter

121. The performance of the play was over. You met Jan backstage. She was crying as she removed her stage makeup. Because she had broken character by waving to you in the middle of the play, because she had said the donkey's lines (Hee Haw) instead of Hermia's, because she had hit Helena, the director had done two things. He had fined her, and he had fired her.

"Let's go walking," Jan said to you.

Often the best thing that a friend (like you) can do is just to be there with the one who is hurting. Just listen. No need to offer advice or fix things.

As you walked in the night, Jan poured out her heart to you.* She liked to sing and dance. She liked the applause. She liked the money she made at Ivy's. She had success as a donkey but utterly failed as Hermia.

She realized that she had both success and failure mixed together.

Without your saying a word, some of the sting of being fined and fired started to fade.

Unlike some 5½-year-olds who will remain unnamed (👤), Jan was getting hungry. It was 9:30 p.m. and she hadn't eaten since lunch.

There was a new sign at Ivy's. "Perfect," said Jan, "but I don't have any money."

"This will be my treat," you told her.

If Ivy's were only open 23 hours each day, what percent of the time would she be open. Round your answer to the nearest percent.

---

* To *pour out your heart* is a metaphor. To *have a breaking heart* is a metaphor. These are not literally true. The only thing a heart pours out is blood—and that goes into the arteries.

Chapter Twenty-three   Blood

Second part: the 𝔐ixed 𝔅ag: a variety of problems from this chapter and previous chapters

262. Your blood starts at the heart, goes through the arteries and then into the tiny capillaries and delivers two good things to the cells. Name them.

330. The happy cells take in those two good things and deposit back into the blood stream some carbon dioxide. Carbon dioxide came from "burning" the food with the oxygen inside the cell.
    The blood went from your heart to your leg muscles through your arteries. It is transported back to your heart through your _____.
fill in one word

464. Your blood circulates: heart → arteries → capillaries (where the cells are) → veins → heart.
    At the cells, food is dropped off.
Where was the food picked up in the first place? (Hint: In Chapter 18 we called it the 𝔓𝔩𝔞𝔠𝔢 𝔬𝔣 𝔉𝔦𝔫𝔞𝔩 𝔇𝔦𝔤𝔢𝔰𝔱𝔦𝔬𝔫 or the 𝔚𝔢𝔩𝔠𝔬𝔪𝔢 𝔆𝔢𝔫𝔱𝔢𝔯.)

626. At the cells, oxygen is dropped off. Where in the circulatory system was the oxygen picked up?

754. Ivy said, "Hi. What would you like to have?"
    Jan's life revolved around makeup. When she looked at Ivy, she first thought that Ivy had done a poor job "doing" her face.
    Ivy explained, "Since I have changed the store to a 24-hour store, I haven't been getting much sleep."
    Your mind was working much faster than either Jan's or Ivy's. The solution was obvious to you. You made the suggestion. Jan and Ivy both loved it.
    Jan became the assistant manager of Ivy's Ice Cream. A 12-hour shift. $26 per hour. The government would take out 20% for income taxes, 6% for medical insurance, 5% for Social Security, and 7% for unemployment insurance.
    How much would Jan get per day?

Chapter Twenty-four
Staying Alive

First part: Problems from this chapter

155. Ivy now had a 12-hour work day (instead of a 24-hour work day). If she slept for 8 hours, what percent of each day would be free? (Translation: What percent of the day would she not be either working or sleeping?)

223. Most people waste a lot of time in their lives. Ivy wasn't one of those. With four hours of free time everyday, she devoted one hour of each day for what she called her "special project."

One hour a day devoted to your special project can change your life. Ivy had always wanted to learn Greek. She had always wanted to be able to read the ancient Greek plays and the New Testament in their original language. (Ελληνικά is the Greek word for *Greek*.)

Keeping it up (← important!) and it wouldn't be that many months before Greek would become a part of Ivy.

Possible special projects: ★ Jogging
 ★ Learning about real estate investments
 ★ Playing the piano

Your turn. Name five other special projects that a person might have. Be creative. Don't just write ★ Running or
 ★ Learning about investments or
 ★ Playing the flute.

Intermission

I, your author, have a special project. It is writing *Life of Fred* books. (You could have guessed that.) For the last 19 years I have gotten up at 4 a.m. and have spent two hours each day discovering what Fred and his friends are doing. The book you are holding in your hands is my 59th book.

Having a special project does work.

Chapter Twenty-four   Staying Alive

Second part: the 𝔐ixed 𝔅ag: a variety of problems from this chapter and previous chapters

341. Staying healthy is one step above just staying alive (which is the title of this chapter).

    If you just stay alive till you are 80, the last 20 years of your life might be filled with aches and pains as you sit in front of the television or play bingo.

    If you stay healthy till you are 80, and can still play tennis or chase your grandkids around the playground, you will be happy that you did the right things during the first 60 years of your life.

    Your daily activities can affect your health. Put an "H" (for healthy) or a "U" (for unhealthy) in front of each of these:

    ____ smoking
    ____ driving a motorcycle at night in the middle of a snow storm
    ____ being a mathematician
    ____ eating a lot of candy
    ____ exercising
    ____ wearing a seatbelt

467. Your mother's sister has short eyelashes. Your mother's mother has short eyelashes. Could your mother's father have a genotype of SL?

624. (continuing the previous problem) If your mother has long eyelashes, what can you say about your grandfather's *phenotype*?

    (Phenotype is the observable trait. Genotype is what genes you have. Your genotype of SS, SL, or LL will determine your phenotype.)

757. Jan started her work as assistant manager at Ivy's Ice Cream. A full container (5 gallons) of 𝒞𝓁𝑜𝓊𝒹𝓎 𝒱𝒶𝓃𝒾𝓁𝓁𝒶 weighs 36 pounds. Two and a half quarts had already been scooped out. How many gallons and quarts were left? (1 gallon = 4 quarts)

888. A gallon is 231 cubic inches. A 5-gallon container (a cylinder) of 𝒞𝓁𝑜𝓊𝒹𝓎 𝒱𝒶𝓃𝒾𝓁𝓁𝒶 has a radius of 6 inches. To the nearest inch, how tall is that container? (Use 3 for π.)

Chapter Twenty-five
Solving Algebraic Equations

First part: Problems from this chapter

158. How many terms in each of these expressions?
    6y + 32 – 20 ____
    $5x^{23}yz^4$ ____
    5 + 7 ____
    (3x + 6)(444) ____

256. How many terms in each of these expressions?
    98967x – x ____
    $5 + 4(x^2 – y^2)$ ____
    $(6 + 6 + 6)^8$ ____

347. Combine like terms in the equation $4y^2 – y^2 + 17 = 3x^2y + 18xy$

435. x squared is x^2.
    x cubed is x^3.
    What is forty-four x cubed times y to the seventh power?

596. You have the same number of nickels, dimes, and quarters in your pocket.
    Let x = the number of nickels you have.
A) How many dimes do you have?
B) How many quarters do you have?
C) How many cents are your nickels worth?
    (If you had 7 nickels, they would be worth 7(5).
    If you had 8 nickels, they would be worth 8(5).
    You have x nickels.)
D) How many cents are your dimes worth?
E) How many cents are your quarters worth?
F) How many cents are your nickels, dimes, and quarters worth?

Chapter Twenty-five   Solving Algebraic Equations

Second part: the 𝕸ixed 𝕭ag: a variety of problems from this chapter and previous chapters

724. Joe, who is known for his big appetite, walked into Ivy's Ice Cream when Jan was on duty. He asked for one scoop of Ivy's famous SHIVERING STRAWBERRY.

Jan was busy with another customer and told Joe to serve himself. This was a mistake. A full 5-gallon container of SHIVERING STRAWBERRY weighs 37 pounds. Joe put the whole thing on a cone. It was the biggest "one scoop" that the world had ever seen.

70% of it melted and ran down his arm before Joe could finish it. How many pounds of SHIVERING STRAWBERRY did he eat?

774. Joe met Darlene outside the store and told her that Ivy's one-scoop ice cream cones are the best in the whole world.

Darlene took Joe's hand, and they walked back into the store. She asked Jan for a one-scoop of *Matrimonial Maple*.

Jan scooped out a ball (a sphere) of *Matrimonial Maple* and gave it to her. It had a radius of three-fifths of an inch. What was the volume of that ball of ice cream?

$$V_{sphere} = \frac{4}{3}\pi r^3$$ (Use 3 for π in this problem.)

857. (continuing the previous problem) Darlene ate three-eights of it and said that she was full. She gave the rest of it to Joe who eagerly devoured it. How many cubic inches of *Matrimonial Maple* did Joe eat?

960. At this point Joe went beyond shivering. He experienced hypothermia. (*hypo* = less than normal; *therm* = deals with temperature. You know the words *thermostat, thermos,* and *thermometer.*)

Darlene offered to hug him to get him warmer. Joe said he didn't do that kind of thing.

Jan offered him 1½ cups of hot milk. Joe drank half of that. (His tummy was getting pretty full.)

He jumped up and down to try to get warm. He become the world's largest milkshake.

How much milk did Joe drink?

Chapter Twenty-six
The Second Step in Solving Equations

First part: Problems from this chapter

159. A scoop of Ivy's Ice Cream cost 47¢.
 In the first hour that Jan worked she sold one scoop to Joe, one scoop to Darlene, and scoops to other customers. She also sold a 17-cent polar bear sticker to a girl for her sticker collection. The polar bear reminded her of cold ice cream.
 These sales totaled $13.80. How many scoops did Jan sell?
 Hints: Always start by "Let x = . . . " and if we say 1380¢ instead of $13.80, we can avoid decimals.

261. The girl resold her 17¢ polar bear sticker to her friend for $3. What percent gain did she have? (Round your answer to the nearest percent.) Hint: The answer is *not* 1765% or 1764% or 17% or 18%.)

377. Joe was full. Really full. He told Darlene, "I won't be able to eat anything until lunch." It was 10:47. How long will it be until noon?

465. (continuing the previous problem) Joe never listens to himself when he is talking. He also doesn't listen to others when they are talking.
 Joe had said, "I'm really full."
 Darlene told him, "Just stop eating until noon."
 Jan offered him a tiny taste of their **Manly Mango** ice cream. It was a very bright orange. Without thinking, he took a little lick.
 If this were a cartoon book, Joe would have exploded. His guts would have been everywhere. But the *Life of Fred* books always stay close to reality and decorum.
 Joe moaned. He told Darlene that he felt really, really bad. Jan suggested that Darlene drive him to her favorite hospital (Beth's Vet). She drove at 12 mph, and they spent 5 minutes in the parking lot trying to find a space. It was now 11:08. How long were they in the car?

Chapter Twenty-six   The Second Step in Solving Equations

Second part: the 𝔐ixed 𝔅ag: a variety of problems from this chapter and previous chapters

576. (continuing the previous problem) How long did they drive before they got to the parking lot of Beth's Vet? Give you answer in minutes and then convert that to hours.

669. (continuing the previous problems) How far did Darlene drive to get to the hospital?

759. As they got out of the car, nature helped the healing process. Joe threw up. He had weighed 174 pounds, 4 ounces. He now weighed 170 pounds, 7 ounces. How much did he lose? (1 pound = 16 ounces)

802. Change $\frac{21}{24}$ into a percent.

870. Solve $8y + 31 = 5y + 82$

920. Darlene looked at the sign outside the hospital. She realized that Beth's Vet was a veterinary hospital. *Veterinary* has five syllables.

Darlene wondered why Jan had suggested that they go there. Later, when she asked Jan, Jan said, "I went there when I was a donkey." That didn't make any sense to Darlene.

Darlene wiped off Joe's mouth, and they got back into the car. She realized that if she married Joe, she would have to be taking care of him for most of their married life. He always seemed to be in trouble.

In the car he rested his head on her shoulder and fell asleep. He was dreaming about ice cream.

Joe's dream:
I was in a room with Ivy's one thousand flavors of ice cream. I invented three new flavors that she didn't have. Boatman's Blueberry. Fisherman's Frangipani. Lazyman's Licorice. This 1003rd flavor is for really lazy people. It is pre-chewed.

Is 1003rd a cardinal number?

Chapter Twenty-seven
Words to Equations

First part: Problems from this chapter

161. $9w - 3 = 4w$

237. $2x + 5 + 4x = 15x$

345. Joe woke up. He told Darlene that he had dreamed that a pelican had been chasing him.

In his dream he was more handsome than in real life.

Joe had a 100-foot head start.
Joe was running at 14 feet per second.
The pelican was running at 16 feet per second.
Let t = the number of seconds before the pelican catches Joe.
How far did Joe run before the pelican caught him?

374. How far did the pelican run before it caught Joe?

380. How much farther did the pelican run than Joe before it caught Joe?

500. Looking at the results of the previous three questions, write the equation.

549. Solve the equation.

Chapter Twenty-seven   Words to Equations

Second part: the 𝔐ixed 𝔅ag: a variety of problems from this chapter and previous chapters

625. Darlene comforted Joe, "Poor dear. Nobody's chasing you. You just had a bad dream."

Joe had forgotten that they had been at Pelican Brother's Jewelry store a couple of weeks ago, where Darlene had pointed out to Joe the lovely wedding ring she liked.

Any amateur psychotherapist could figure out who the pelican was in Joe's dream.

Darlene wasn't paying attention to her driving. She was thinking of Joe and how cute he was. She was doing 60 mph on a city street.

The cop was $\frac{2}{5}$ of a mile behind her when he turned on his lights. He was going 84 mph. How long before he caught up?

680. The speeding ticket was for $500. The speed limit was 35 mph. The fine was $x for each mph over the speed limit plus court costs of $75. Find the value of x.

801. When Darlene works, she makes $8 an hour. How long would it take her to earn $500? (Use a conversion factor.)

856. Darlene once briefly worked at Ivy's Ice Cream. In an hour she could fill order forms that made a stack that was $3\frac{7}{8}$ feet tall. In an hour Ivy could do $8\frac{1}{4}$ feet.

Darlene wasn't a very good worker. She took time off to do her nails and to call her girlfriends and read wedding magazines.

How much larger is $8\frac{1}{4}$ than $3\frac{7}{8}$?

Chapter Twenty-eight
Breathing

First part: Problems from this chapter

131. Your heart beats **involuntarily**. You don't beat your heart. It beats without your having to think about it. You don't have to stay up at night and tell your heart to beat–beat–beat.

By contrast, you tell your muscles to run and jump. These are **voluntary** actions. You decide whether you are going to dance or sing.

You are at Stanthony's PieOne pizza place. In front of you is his famous combination pizza with 82 toppings. You pick up a piece and put it into your mouth (voluntary action), chew and swallow it (voluntary action). Is digestion a voluntary or involuntary action?

275. (trickier question) Is blinking your eyes voluntary or involuntary?

372. Is breathing voluntary or involuntary?

436. Jan finished her 12-hour shift at Ivy's. She had $193.44 in her pocket. (We did that computation back in Chapter 23.)

She had twelve hours before she would have to be at work again. Subtracting eight hours for sleep, she knew that she had four hours for . . . **shopping!** Forget about saving money to pay the rent. Even the thoughts of being in the theater were drifting away. She had money. She needed to spend it.*

With 240 minutes (4 hours × 60 minutes/hour) to shop, how long would her money last if she spent it at the rate of $5 every 4 minutes?

Give your answer to the nearest minute. I'm going to use a conversion factor.

---

* Don't write *She had money, she needed to spend it.* That's called a run on sentence—two sentences smashed together with a comma.

Any of these are correct:
> *She had money. She needed to spend it.*
> *She had money; she needed to spend it.*
> *She had money, and she needed to spend it.*
> *Since she had money, she needed to spend it.*

Chapter Twenty-eight   Breathing

Second part: the 𝔐ixed 𝔅ag: a variety of problems from this chapter and previous chapters

552. Jan had 4 hours of free time. She spent 2 hours and 35 minutes shopping. How much time was left?

601. One of Jan's purchases was $9\frac{3}{8}$ pounds of sparklettes. She wasn't sure what they were, but they were cute.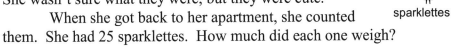
When she got back to her apartment, she counted them. She had 25 sparklettes. How much did each one weigh?

701. If her 25 sparklettes cost a total of $4.25, how much did each one cost? Give your answer in cents.

803. (continuing the previous two problems) How much would a ton of sparklettes cost? (1 ton = 2,000 pounds)

921. Three-eights is what percent of $\frac{15}{16}$?

69

Chapter Twenty-nine
Still Breathing

First part: Problems from this chapter

160. Did you ever notice how many people and how many animals go around breathing? They just keep on sucking in oxygen (20% in) and  breathing out less oxygen (16%). If they kept doing that, you would think that they would use up all the oxygen on the planet. Then, of course, all the people and all the animals  would die. But people and animals have been around for hundreds of years—some say even longer—and we haven't run out of oxygen. Why?

258. When kids first learn about the circulatory system—how the blood travels around in the body—they think it looks like this: The heart pumps out the blood. It visits the lungs to get oxygen. It visits the small intestine to get the food. It delivers the oxygen and food to the cells. The cells are delighted with these gifts.

That picture is *almost* right. The only problem is that this is a **HIGH PRESSURE** system. When doctors take your blood pressure (by putting a blood pressure cuff on your arm), they might get a reading of approximately 120/80. The 120 is when the heart is pushing. The 80 is between pushes.

The difficulty is with the lungs. Oxygen goes into your lungs when you breathe in. The blood needs to pick up that oxygen. If you had regular blood vessels in your lungs, the oxygen couldn't get into your blood.

Did you ever notice blood vessels in your arms? You can't breathe through those blood vessels—even if you wave your arms around. Those

Chapter Twenty-nine Still Breathing

blood vessels can withstand the **HIGH PRESSURE**. The **HIGH PRESSURE** is needed to push your blood all the way to your feet and back again.

In your lungs the blood vessels are very thin. Oxygen passes easily from the air into your blood. If the blood pressure were 120, like the rest of the body, those thin lung blood vessels would pop.* Your lungs will fill with blood, which would be uncomfortable for a couple of minutes.**

THE Health Caboodle

The Official Newspaper of Healthy Living   10¢

Lungs Threaten to Go on Strike!

KANSAS: Lungs held a press conference today. They announced that they will not tolerate the high blood pressure that the rest of the body likes.

True Lung
file photo

"We demand that we get blood delivered to us that is meet, right, and salutary for us."

The reporters ran to their dictionaries. They found that *meet* means fitting and proper. They found that *salutary* means promoting health.

You are still alive. Otherwise, you wouldn't be reading this. Your lungs have a low blood pressure of about 15 (instead of 120). Make a wild guess—how does your body accomplish this trick.

---

* The technical word is *rupture*.

** After that, you wouldn't feel uncomfortable. In fact, you wouldn't mind if they stuck needles in you or gave you extra chores to do. Without oxygen to your brain, you would first pass out and then pass away.

Chapter Twenty-nine   Still Breathing

Second part: the 𝓜ixed 𝓑ag: a variety of problems from this chapter and previous chapters

348. Jan looked at her 25 sparklettes. Sixty percent of them were red. Thirty-two percent of them were green. The rest were yellow. How many were yellow?

438. Joe doesn't exercise very much. His diet is not very good. (This is an example of litotes—saying the opposite of what is true and adding a *not*.) The truth is that Joe's diet stinks. His most recent blood pressure reading was 139/87. What is his systolic reading?

577. The set of Jan's favorite activities = {dancing, singing, shopping, napping}.
    The set of Joe's favorite activities = {fishing, eating, napping}.
Are these sets disjoint?

681. When Jan was memorizing lines for a play, she could learn 6 lines in 8 minutes. At that rate how many lines could she memorize in 44 minutes. Use a conversion factor.

782. $(7.8)^2$ = ? Use pencil and paper rather than a calculator.

805. Jan had started her shopping spree with $193.44 in her pocket. She spent $4.25 on sparklettes.   She spent $150.83 on cat food. She spent the rest on happy gingerbread men.

What percent of her original $193.44 did she spend on these happy men?
Give your answer to the nearest percent.

Chapter Thirty
How Oxygen Is Carried in the Blood

First part: Problems from this chapter

188. Jan's poor management of her money was **an albatross around her neck**. Is this a simile or a metaphor?

226. The blood that comes out of your heart and through your aorta is at high pressure—around 120.
    It has a long way to travel. It begins in your arteries (which carry blood away from your heart*) and then to the capillaries to all the cells in your body (including your bones).
    On the way back to your heart the blood travels through your veins at a lower pressure.
    Now if you were designing a human body—let's say the blood vessels in an arm—would you put the arteries near the skin and the veins on the inside or the other way round?

338. Scientists nowadays are finding new ways to design our bodies. There are artificial hearts, plastic lenses to replace the yellowed lenses in the eye that older people sometimes get (cataract surgery), and false teeth.
    What do you think of this idea? Let's transfer eyeballs to the palms of our hands. That way we can look at the back of our heads to make sure that our hair isn't sticking up. That way we can look at our own faces directly when we pick our noses.

---

* <u>A</u>rteries and <u>a</u>way is a nice mnemonic.
And you know what I, your reader, am going to ask!
Yes. A mnemonic (ni-MON-ik) is a trick to help memorize something.
    It's hard to remember how to spell the word *mnemonic* because of the silent *m* at the beginning of the word. My mnemonic for remembering that is *mnemonic* and *memory* both start with an *m*.

Chapter Thirty   How Oxygen Is Carried in the Blood

Second part: the 𝔐ixed 𝔅ag: a variety of problems from this chapter and previous chapters

441. Hemoglobin molecules weigh anywhere from 17,000 to several million. The little iron atom weighs 56. What percent of the smallest hemoglobin (17,000) molecules is that little iron atom?
   Round your answer to the nearest hundredth of a percent.

502. When Jan got back to her apartment, she put her sparklettes and her happy gingerbread men on the kitchen table and headed to her television set, which was 14 feet away. She was tired. She walked at 3 feet per second. Using fractions, find out how long she walked.

639. She turned on her favorite show: H͓o͓u͓s͓e͓s͓ o͓f͓ F͓a͓m͓o͓u͓s͓ M͓o͓v͓i͓e͓ S͓t͓a͓r͓s͓. She imagined that if she saw this, she could learn how to become a movie star.*

The show was a little bit more exciting than watching people go fishing. This is the house of Alfredo von Futz. It has a kitchen, a living room, a bathroom, and a closet.
   This is the house of Blush Bubbles. It has a kitchen, a living room, a bathroom, and a closet.

Every time the show introduced a new house, Jan reached over, grabbed a chocolate doughnut and ate it. Nine houses, nine doughnuts. She washed them down with 284 Calories of Sluice. The doughnuts and the Sluice were 6,800 Calories. How many Calories were in each doughnut?

783. The natural numbers = {1, 2, 3, 4, 5, . . .}. If y is a natural number, how many solutions would $3y + 7 < 20$ have?

---

* And if you watch a lot of horse racing, you can learn how to be a horse.

Chapter Thirty-one
Large Numbers

First part: Problems from this chapter

134. One neon atom has an atomic weight of 20.18.
What is the atomic weight of a dozen neon atoms?

227. What is the weight (in grams) of a mole of neon?

339. What is the weight (in grams) of an Avogadro's number of atoms of neon? This is the same as asking how much 6.0221236×10^{23} atoms of neon weigh.

485. How much (in grams) does a dozen neon atoms weigh?

555. There are roughly seven billion people on earth. To make the arithmetic easier, suppose Sally had asked for seven quadrillion dollars instead of just one quadrillion dollars.
　　If she distributed that evenly to the seven billion people, how much would each receive?

682. In the *Your Turn to Play*, we found that the molecular weight of glucose, $C_6H_{12}O_6$, is $6\times12 + 12\times1 + 6\times16 = 180$.
(The atomic weight of C is 12; of H is 1; of O is 16.)

　　A mole (6.0221236×10^{23} molecules) of glucose weighs 180 grams.
　　At the dinner table if your mom asks you to pass her a mole of glucose, how many ounces of sugar would that be?
　　(One gram is about 0.035 ounces.)

Chapter Thirty-one   Large Numbers

Second part: the 𝔐ixed 𝔅ag: a variety of problems from this chapter and previous chapters

722. On the Houses of Famous Movie Stars television show, there was a purple house in Los Angeles that Alfredo von Futz moved into. He bought it for $360,000. After he moved into it, it became a "house of a famous movie star." It was now worth $500,000.

To the nearest percent, what percent increase was that?

790. Jan practiced putting on her makeup as she watched the television show. She told herself, "You never know when they might knock on your door and ask you to be in a movie."

She ignored the fact that she was in Kansas rather than in Hollywood where many movies are made. She ignored the fact that only Ivy and a couple of close friends knew where she lived.

"Maybe if I paint my apartment purple like Alfredo's house, they will know I'm available to star in a movie."

Jan wasn't sure who *they* were.

Suppose there was a 0.043% chance that *they* would knock on Jan's door. Round that to the nearest percent.

806. Jan thought of a new way to put on lipstick. She grabbed one of her chocolate doughnuts and smeared it with Hollywood Pink lipstick. Then she ate the doughnut. Her lips were now pink.

So were her teeth.

There was a knock on the door. It was a little girl with a camera. The girl said, "Hi. I'm from Hollywood and I'm looking for a woman to star in the movie I'm producing."

They were knocking on Jan's door! For every million knocks on her door, 0.043% of them would be Hollywood producers. How many is that?

871. If Jan received 4 knocks on her door every week, how long would it take to receive a million knocks?

932. (harder question—just for advanced students) How long might Jan expect to wait to receive a knock on the door from a movie producer?

Chapter Thirty-two
Stoichiometry

First part: Problems from this chapter

"Ha! Ha! Ha!" said the little girl with a camera. "Did you really believe me when I said I'm from Hollywood? That only happens in your dreams."

Jan had fainted because of the shock. Her mouth was open and her pink teeth showed.

While Jan is passed out, let's balance some chem equations. Please remember the two Giant Hints:
#1 Start by balancing the letter that appears in the fewest molecules in the equation.
#2 Start by looking at the most complex molecule first.

162. H_2O_2 is hydrogen peroxide. You can buy it in drug stores. You can put it on wounds or use it to brush your teeth.

It changes into water and oxygen. The oxygen helps disinfect the wound and helps whiten teeth.

Balance the skeleton equation $H_2O_2 \rightarrow H_2O + O_2$

225. The little girl with the camera brushed Jan's pink teeth with hydrogen peroxide. When Jan awakens, her teeth will be white again.

Carbon monoxide combined with oxygen will produce carbon dioxide.

Balance the skeleton equation $CO + O_2 \rightarrow CO_2$

333. Balance $NH_4NO_3 \rightarrow N_2O + H_2O$

469. Balance $H_2 + N_2 \rightarrow NH_3$

506. Some houses have propane furnaces. Burning propane is given by the skeleton equation $C_3H_8 + O_2 \rightarrow CO_2 + H_2O$. Balance that equation.

Chapter Thirty-two   Stoichiometry

Second part: the 𝔐ixed 𝔅ag: a variety of problems from this chapter and previous chapters

594. Jan woke up. The bad news was that the little girl wasn't a movie producer who would put Jan in the movies. The good news is that Jan's teeth were white.

While Jan was unconscious, the little girl had eaten some of Jan's chocolate doughnuts. Jan's stash of 6 pounds of doughnuts was now 3 ounces lighter. (1 pound = 16 ounces). How much was left?

641. "Why are you here?" Jan asked.

"I like to take pictures," she explained. "Look at my camera. I will take 21 glamour pictures of you and toss in a 79¢ frame. The whole thing will be $11."

How much would each picture be? (Hint: Work either in cents or in dollars.)

702. Potassium chlorate decomposes into potassium chloride and oxygen.

The chemical symbol for potassium is K, because the symbol for phosphorous is P.

The chemical symbol for chlorine is Cl, because the symbol for carbon is C.)

$$\text{Balance } KClO_3 \rightarrow KCl + O_2$$

794. Jan wanted the little girl's camera. She asked her how much was the camera. The little girl thought: *I bought it for $46 and would like to make a 20% profit if I sold it.* How much was the little girl willing to sell her camera for?

891. If the little girl sold her camera for $62, what would be her percent profit? (Round to the nearest percent.)

Chapter Thirty-three
Small Numbers

First part: Problems from this chapter

179. With all this excitement Jan had the tiniest bit of sweat on her forehead. Sweat is mostly water with a little bit of salt (and other stuff).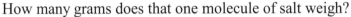

Salt is sodium chloride, NaCl. We already know that Cl is the chemical symbol for chlorine. Na is the chemical symbol for sodium.*

The atomic weight of Na (sodium) is 23. The atomic weight of Cl (chlorine) is 35.**

How many grams does that one molecule of salt weigh?

378. The burning of CH_3NH_2 is given by the skeleton equation

$$CH_3NH_2 + O_2 \rightarrow CO_2 + H_2O + N_2 \quad \text{Balance this equation.}$$

(This isn't super easy. ← litotes)

---

* Na is the chemical symbol for sodium because N is the chemical symbol for nitrogen.

    In calculus we are going to get even sillier when we explain that
- m is the symbol for slope because
- s is the symbol for arc length because
- a is the symbol for acceleration.

There are more concepts than there are 26 letters in the alphabet.

** I'm keeping the numbers simple. Chlorine, for example, has an atomic weight of about 35.453 because of its various isotopes that occur in nature.

Hey! I, your reader, need to know: What's an isotope?

Isotopes (ICE-oh-topes) of chlorine are atoms of chlorine with different numbers of neutrons in them. About 76% of chlorine atoms have 18 neutrons and about 24% of them have 20 neutrons in them.

And . . . is it too much to ask? . . . What's a neutron?

You have asked two questions. The answer to the first one is yes. Otherwise, this footnote could go on forever.

Chapter Thirty-three Small Numbers

Second part: the 𝕸ixed 𝔅ag: a variety of problems from this chapter and previous chapters

489. Name a number smaller than $\dfrac{\pi}{890{,}385{,}227.614}$

530. When Joe was 18, he lost his fishing rod in a bet with C.C. Coalback.

Coalback challenged Joe to a **Name the Largest Number Contest**. Coalback said he would be polite and let Joe go first.

Joe thought for a moment and said 90979379239908094300347.

Coalback smiled and shouted 90979379239908094300348. Coalback now owned Joe's fishing rod.

They ran the contest a second time. Joe said a billion, billion, billion. Coalback said a billion, billion, billion *and one*.

Joe wanted his fishing rod back. He offered to bet his fishing boat against Coalback's fishing rod. The contest would be to **Name the Smallest Positive Number**.

Again, Coalback was polite and let Joe go first. Joe thought for a moment and said 0.0000000000000000000000000001.

Coalback said 0.00000000000000000000000000005 and won Joe's fishing boat.

⭐ There is no largest number, because you can always add one to get a larger number.

⭐ There is no positive number that is right next to zero, because you can always divide that number in half and get a number even closer to zero.

⭐ Lots of people nowadays like to say "perfect." There is nothing you that do that is ever "perfect." Spending a couple more hours will make your work a little closer to that ideal of perfect.

You can be more graceful. You can be more skillful at playing tennis. But you can never be "more perfect" or "more unique."

Perfect and *unique* are the end of the line. They are called **absolute adjectives**. For all those who think English is easier than math, here is your question: Name some more absolute adjectives.

Chapter Thirty-four
Shortcuts

First part: Problems from this chapter

110. The biggest shortcut that many algebra students like to take is leaping directly from the English to the equation.

Let's suppose that Jan spotted a real movie producer. He was 60 feet away from her. He also spotted her. He was running at 9 feet per second *away* from her.

She, of course, was chasing him. She was running at 12 feet per second. We want to know how long it would be before she caught him.

The English-to-equation leapers immediately write:

$12t + 9t = 60$ or
$\qquad 9t + 60 = 12t$ or
$\qquad\qquad 21t = 60$ or
$\qquad\qquad\qquad t = 60 + 9t + 12t$ or
$\qquad\qquad\qquad\qquad 60t = 9 + 12$.

And those leapers always seem to ask, "Why didn't I get the right answer?"

What is the first thing that should be written down?

231. How far did Jan run?
263. How far did the movie producer run?
301. How much farther did Jan run than the movie producer?

The English-to-equation leapers complain, "It takes so much work to write down Let t = the number of seconds that they ran.
     Then 12t = the number of feet that Jan ran.

Chapter Thirty-four Shortcuts

Then 9t = the number of feet that the movie producer ran. Now really... is it <u>that</u> much work to write that down?

Then they complain, "But it takes too much thought to figure all that out."

And they are right—it does take more work to think—even more work than just writing three lines.

Reality Time

In the old days men could get jobs digging ditches. If you had a strong back, they would hand you a shovel and you would dig and make money.

That's no longer true. Machines can dig faster than you can. And they are cheaper than hiring you.

In the not-so-old days if you didn't have a decent education (and that includes algebra!) you could always work in a warehouse or be a cashier or wait on tables in a restaurant.

Those jobs are disappearing.

robot cashier

If you owned El Quacko Taco, would you rather hire someone at $12 an hour or get a machine that...

- ✓ is never late
- ✓ doesn't complain
- ✓ is always accurate
- ✓ doesn't need health insurance
- ✓ works for pennies per hour

The **reality** is that our world is changing... fast.

At the very minimum, get an education that will allow you to repair those machines.

Better yet, get an education—say in robotic engineering—that will allow you to design those machines.

Chapter Thirty-four Shortcuts

Second part: the 𝔐ixed 𝔅ag: a variety of problems from this chapter and previous chapters

424. In twenty seconds Jan caught up with the movie producer. She could tell just by looking that producing movies was his life.*

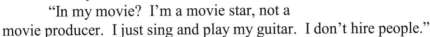

He was out of breath having run for 20 seconds in an attempt to elude her. "I know—you want my autograph," he panted.

"Let's both sign," Jan said, "our signatures on a movie contract for me. I'm sure I will be wonderful in your movie."

"In my movie? I'm a movie star, not a movie producer. I just sing and play my guitar. I don't hire people."

Twice, Jan had imagined that she had seen someone who would get her into the movies. Twice, she had been wrong.

What's two times zero?

528. In calculus, how will you write the expression for the area under the curve y = log x from x = 3 to x = 5?

595. $6\frac{2}{5} \div 2\frac{2}{7}$

700. Jan was embarrassed. She offered the movie star five apologies (all of equal length) and then said goodbye, which took 4 more seconds. Her 5 apologies and her goodbye took a total of 35 seconds. How many seconds did each apology take?

The first commandment of doing word problems:

You will start by writing: Let t = the number of seconds for each of Jan's apologies.

---

* Of course, she had also thought that the little girl with a camera was a movie producer.

Chapter Thirty-five
Reducing Fractions

First part: Problems from this chapter

135. Reduce this fraction $\dfrac{31688113005897}{31688113005897}$

235. It was time for Jan to head back to Ivy's Ice Cream to begin her 12-hour shift. She said to herself, "Maybe I'll meet a movie producer while I'm scooping ice cream. That would be ▒▒▒▒▒."*

When Jan got there, she found a pile of 19,005 ice cream cones and a note from Ivy asking her to stack them up evenly.
   A) Could she divide them into two even piles?
   B) Could she divide them into three even piles?
   C) Could she divide them into five even piles?

346. We know that we can reduce $\dfrac{670}{8330}$ to $\dfrac{67}{833}$ by dividing top and bottom by 10.

Dividing by 10 might look like $\dfrac{67\cancel{0}}{833\cancel{0}}$

What about reducing $\dfrac{12}{732}$ by doing $\dfrac{\cancel{1}2}{73\cancel{2}}$ and getting $\dfrac{1}{73}$?

470. In the ice cream store business you meet all kinds of people. Jan's first customer of the day didn't look like a policeman or a carpenter or your father.

He said, "I'll have $\dfrac{1}{3}$ of a scoop of chocolate, and $\dfrac{1}{5}$ of a scoop of 𝖺𝖭𝖈𝗁𝗈𝖵𝖸-𝖫𝖺𝖬𝖡 𝖶𝗂𝗍𝗁 𝖺 𝖱𝗂𝖡𝖡𝗈𝖭 𝗈𝖥 𝖸𝖺𝖬 and the rest of the scoop will be vanilla. Jan wanted to scream, "Don't clown around!" but she just filled the order. How much vanilla did Jan scoop?

---

* You have to fill in that word. It depends on how old you are. If you are really old, you would say, "That would be *swell*." A little younger, "That would be *neat*." Younger yet, *cool*. Youngest, you use the absolute adjective and say, "That would be *perfect*."

Chapter Thirty-five Reducing Fractions

Second part: the 𝔐ixed 𝔅ag: a variety of problems from this chapter and previous chapters

578. In the previous chapter (in the answer to problem #424) we learned that log 10^3 is 3 and log 10^7 is 7 and log 10^π is π. You don't quite know what log means, but you do know log 10^{47} equals 47.

If you have a scientific calculator* (the kind that has sin, cos, tan, and log keys), you can easily find log 4.

You just hit ⎡4⎤ and then ⎡Log⎤ and then ⎡=⎤. Easy peasy.**

Out will pop something like 0.602 (on cheap calculator).
Or 0.60206 (on a regular calculator).
Or 0.60205999132796239042747778944899 (on my silly calculator).

You know what 10^7 means. It's 10 × 10 × 10 × 10 × 10 × 10 × 10. Now I want you to tell me what $10^{0.60206}$ is approximately equal to.

I know that $10^{0.60206}$ doesn't make any sense to you (yet), but you can still find the answer. Ask this question of your older sister who is studying advanced algebra. If she's an 𝒜 student, she should know the answer.

671. The clown said to Jan, "You know that you have to deal with a lot of stress if you are in the movies. Too many people who want to act in movies can't take it. They go nuts when they hit the slightest difficulty."

"How would you know? You are only a clown," Jan said.

"Can you deal with stress?" he asked. "What's seven divided by one-millionth?" Help Jan get the answer (and then we'll find out why the clown was asking that question.)

---

* Your basic arithmetic calculator has +, −, ×, and ÷ and maybe √ .
A scientific calculator is the last one you will ever have to buy. They cost around $10—and in a couple of years from now be given away for free in cereal packages. Prices of things that do computing have been dropping like crazy.

** Easy peasy means the same as a piece of cake or easy as pie or it happens automagically or it's low-hanging fruit or it's no sweat.

Chapter Thirty-six
Division by Zero

First part: Problems from this chapter

189. Jan was getting a little stressed out by this clown with his weird ice cream order and his stupid arithmetic question. But she didn't want to be fired by Ivy for being rude to a customer. She just smiled.

She was glad he didn't ask her advanced algebra questions such as $\log 10^{72} = ?$"*

He smiled. "You say that I'm only a clown. Because I dress as a clown, does that mean that I am only a clown?

A) If you put feathers in your hair, does that make you a bird?

B) If you steal things, does that make you a thief?

C) If you know that $\log 10^e = e$, does that make you look smarter?

D) If you are good at acting and singing, does that mean you will get a part in a movie?

230. The man in the clown outfit said, "You know there are two kinds of people that I meet: those who *say* they are good at singing and dancing and those who *are* good at singing and dancing."

Jan's face lit up. "I'm both." She started singing her aNChoVY-LaMB With a RiBBoN oF YaM song and danced around the ice cream store. She was happy to entertain the clown. Jan didn't realize whom she was auditioning for.**

Earlier she had believed that the little girl with a camera was a movie producer because she *said* that she was.

Then she had believed that someone was a movie producer because he *looked like one*—holding a camera and wearing sunglasses.

The man in the clown outfit looked into Jan's eyes and said, "I wear this clown outfit so that people won't know who I

---

* Actually, that's an easy question for us. $\log 10^{72} = 72$. It wasn't for Jan who didn't know anything about logs.

** *Whom.* Auditioning for *whom*. Math is easier. Don't divide by zero.

86

Chapter Thirty-six Division by Zero

really am. I want to see how they react under stress. I want to see how they can sing and dance when they are not trying to impress me.*

"This is your lucky day. We have the script. We have the title: *Gone with the Breeze*. We have the sets built. All we are looking for is the woman who will play the part of a Southern gal named Crimson."

He pulled out a piece of paper, filled in Jan's name, and handed it to her . . .

> ### *Movie Contract*
>
> Movie Producers of Hollywood, Inc. hereby offers to ___*Jan*___ the leading role in *Gone with the Breeze*. She will be paid $6,000 per day with a guaranteed minimum sum of $60,000.
>
> Work to begin next Monday at our studio. An airplane ticket will be provided free of charge.
>
> The fine print ➔ *Lo único en ese cuarto era un gran escritorio antiguo. Fred solía dormir en la parte superior de él. Cuando el janitor supo lo que estaba haciendo Fred, le dio un saco de dormir chiquito. Sólo medía tres pies de largo, pero era el tamaño perfecto para Fred. Guardó el saco de dormir debajo de su escritorio, lo cual hizo que fuera una cuevita segura para Kinyi y él. A Fred le gustaba hablar con Kinyi. Kinyi no decía mucho, pero era un muy buen oyente. Hace años, Fred lo consiguió a Kingie gratis en el restaurante Rey de Papas Fritas. Cuando Kinyi era nuevo, Fred podía apretar su barriga y Kinyi cantaba una cancioncita sobre las papas fritas. Lo único en ese cuarto era un gran escritorio antiguo. Fred solía dormir en la parte superior de él. Cuando el janitor supo lo que estaba haciendo Fred, le dio un saco de dormir chiquito. Sólo medía tres pies de largo, pero era el tamaño perfecto para Fred. Guardó el saco de dormir debajo de su escritorio, lo cual hizo que fuera una cuevita segura para Kinyi y él. A Fred le gustaba hablar con Kinyi. Kinyi no decía mucho, pero era un muy buen oyente. Hace años, Fred lo consiguió a Kingie gratis en el restaurante Rey de Papas Fritas. Cuando Kinyi era nuevo, Fred podía apretar su barriga y Kinyi cantaba una cancioncita sobre las papas fritas.*
>
> Accepted by: _____
> Sign here

Jan's heart was singing. She didn't read the fine print. She just signed.

Your question: Where do you think Jan will be next Monday?

350. Solve $\frac{w}{3} = \frac{7}{8}$

442. In kindergarten you were told that you can't subtract 8 from 3. In algebra we will do that. $3 - 8 = -5$

In beginning algebra you will be told that you can't take the square root of a negative number. In advanced algebra we will do that. $\sqrt{-4} = 2i$

Make a guess. When will we divide by zero?

---

* When two or more paragraphs are quotes from the same speaker, only the last paragraph has quotation marks at the end.

Chapter Thirty-six Division by Zero

Second part: the 𝔐ixed 𝔅ag: a variety of problems from this chapter and previous chapters

529. On which of these can we use cross multiplying?

A) $\dfrac{4}{5} = \dfrac{x}{6}$

B) $\dfrac{2}{3} + \dfrac{4}{x} = \dfrac{7}{8}$

C) $\dfrac{5}{x} = \dfrac{1}{2}$

D) $\dfrac{11x}{100} = \dfrac{5}{7}$

603. When Jan would finish her 12-hour shift, she normally would have been paid $193.44. Because of the $100 she gave to the clown, she would only get $93.44. Is that number evenly divisible by 3 or by 5?

711. How many pennies would you have to subtract from $93.44 in order to get a number that was evenly divisible by both 3 and 5?

789. Two months ago I, your author, read eight non-fiction books. Last month, another eight non-fiction books. This month, I've only read five books (but the month is only 60% done). There is so much I want to know before I die. I read biographies, histories, economics, science, poetry, and yes, even math. I learn more per day now than when I was in college.
    Yesterday, I was reading this.

> In short, the chart below puts the lie to the alleged virtuous circle of Keynesian stimulus. There has been no pump-priming of consumption spending, production, jobs, income, and more of the same.
> Indeed, the Fed's balance sheet has grown by *900%* during the last 16 years—from $500 billion to $4.5 trillion. By contrast, labor hours have risen by only *6.7% on a cumulative basis*.

I almost fell out of my chair. My pet turtle—if I had one—knows more math than this guy. Since I had bought the book, I reached for a pencil and scratched out one of the numbers. Which one?

872. Solve $4x + 6x + 18 = 27$

Chapter Thirty-seven
Bones

First part: Problems from this chapter

136. Newborns have about 350 bones. Adults have 206. What percentage of bones are "lost." Give your answer to the nearest percent.

210. Will a reduction have to be performed on Fred?

278. Can a solipsist feel pain?

355. Match each medical name in the first column with its ordinary name in the second column.

sternum	thighbone
patella	ear
pinna	skull
cranium	kneecap
femur	breastbone

437. A person's femur is about 27% of their height. This is true for both men and women.
 Some anthropologist* was digging in some field in Kansas and found a thighbone of someone who died a zillion years ago. It was 18 inches long. How tall (to the nearest inch) was that person?

---

* An anthropologist is someone who studies anthropology. **That's kinda dumb. I, your reader, now need to know what anthropology is. And don't say it's what anthropologists study!**
 Would I do that? **Yes you would.**
 Anthropology is the study of people. (*anthropo* = people, *ology* = study of) Everything about people: their origins, their physical development, their cultural development, their beliefs.
 It is such a BIG field. If you were a serious anthropologist, you would probably want to specialize in some small part of anthropology.

Chapter Thirty-seven Bones

Second part: the 𝔐ixed 𝔅ag: a variety of problems from this chapter and previous chapters

531. Which of these is/are not bones?
   sternum, cranium, pinna, eyeball, femur, patella

687. A mom came into Ivy's Ice Cream with her four kids. Here are the sets of flavors that each kid ordered:
   Kid #1 = {chocolate, Shivering Strawberry}
   Kid #2 = {Vacant Vanilla, Crunchy Cinnamon}
   Kid #3 = {Shivering Strawberry, Crunchy Cinnamon}
   Kid #4 = {Vacant Vanilla, chocolate, Crunchy Cinnamon}
Is any pair of these sets disjoint?

758. The cardinal numbers are the numbers used to count the members of a set. (**card A** means the cardinal number associated with set A.)
   Looking at the previous problem, is it true that: card #1 = card #2 = card #3 = card #4?

804. Let x = the cost (in cents) of one scoop of ice cream. The cost of a cone is 18¢. The kids had 9 scoops and 4 cones. The total bill was $4.50. How much is the cost of one scoop of ice cream?

838. Jan thought of the perfect new ice cream flavor: Bison-Blueberry.

She phoned Ivy to tell her. She called at 1:20 a.m. Ivy had gone to sleep at 9:50 p.m.

How long had Ivy been sleeping before Jan woke her up?

936. Solipsists will phone others at 1:20 in the morning and think nothing of it. Solipsists think that no one else matters. Others don't feel pain because they don't really exist.
   How much pain is involved in finding the exact value of

$$\frac{78{,}398 \times 32\tfrac{1}{2} \times 0}{88888} \div \frac{5531.007 \times 39\tfrac{2}{3} \times 19}{44^4 - 33^7 + 1{,}000{,}005}$$

Chapter Thirty-eight
Better Bones

First part: Problems from this chapter

163. It was 1:30 in the morning. Men were putting up new signs on the Ivy's Ice Cream building. Ivy was expanding her business. In addition to a thousand flavors of ice cream, she was going to offer almost every kind of pie.

The grand opening would be at nine o'clock this morning. Radio, television, Internet advertising announced: **3 pies and 8 scoops of ice cream for $9.36!**

Jan knew that scoops of ice cream are 42¢. Showing your work and beginning with *Let x = the price of a pie*, find the cost of one of Ivy's pies.

234. Did you see the price of one of Ivy's pies? Jan did and she was amazed.* These pies weren't small; they were regular-sized pies.** She headed to the back of the store where the pies were going to be made. There were 38 50-pound sacks of flour. How many tons is that? (1 ton = 2,000 pounds)

367. Flour is cheap when you buy it in 50-pound sacks. The biggest expense Ivy would face in making pies is the cost of hiring people to mix

---

* When you have a compound sentence (two sentences joined by the conjunctions *and, or,* or *but*), you normally put a comma before the conjunction. That's the rule in English. The exception is when the when the two sentences are short.

English has lots of exceptions.

** You can join two sentences with a conjunction or with a semicolon. You can't join them with a comma, that would be called a run-on sentence. (Did you see what I just did?) One English teacher I had back in high school would flunk any paper that had a run-on sentence in it. She would probably flunk this book.

Chapter Thirty-eight Better Bones

the ingredients, roll out the pie crust, stuff the ingredients into the crust, and put the whole thing into the oven. Then wait 40 minutes and pull it out of the oven and put it where the food servers can get it and take it to the customer's table.

How do you make pies to sell for $2 each? You lower your expenses. Ivy had purchased a completely automated pie-making kitchen. The sacks of flour were carried by robots to the mixing area. The whole process was done by machine, including taking the freshly baked pies out to the customer's tables.

The Debate
Shall We Get Rid of the Automated Pie-Making Machines?

Argument for Yes: These machines are horrible. They will eliminate jobs. Instead of having twelve people working in that kitchen, there will be nobody. People are more important than machines.

Argument for No: Those who want to smash the machines are very nearsighted. They only see the twelve jobs that are lost. They don't see the big picture. Our world is so much better because of the machines we have built.

Do you want more employment? Instead of hiring twelve people in Ivy's kitchen, you will need 14 people if you eliminate the electric mixer and have the workers stir the dough by hand.

You'll need 18 people if you get rid of the electric or gas ovens and have the workers chop wood to burn in a wood stove.

Machines are designed to make work easier.

Can you imagine how much a shirt would cost if we got rid of sewing machines? Imagine factories filled with people who were hand-sewing shirts. Shirts would be much more expensive. You might own three shirts instead of 30.

Throw away your calculators and do everything with pencil and paper. With calculator, 2 ÷ 23 = 0.08695652173913043478260086 . . .

Without calculator, $23 \overline{)2.00000}$
$\underline{1\,8\,4}$
16 etc.

Chapter Thirty-eight Better Bones

Smash the machines and you would be poor and working all day long doing unpleasant work.

❀   ❀   ❀

Life can change in an instant. Three times Jan thought she had met a movie producer. Three times she thought she would be transported in an eye blink from ice cream scooper to Queen of Hollywood.

Jan's thoughts had been on acting, singing, movies, dancing, and theater, but change was going to come from a totally different direction.

Instead, Jan lost her job—no more scooping ice cream.

Jan went into *deep think* . . .

1. Ivy is making lots of money owning a business.
2. I'm not making much trying to stay employed.
3. I'll start a business.
4. Ivy's Ice Cream is successful.
5. I'll start Jan's Jams.

Which of these has Jan failed to consider?

A) Ivy knows a lot about the ice cream business. She has been doing this for years. Jan knows nothing about jams.
B) It takes some saved-up money to open a new business.
C) Is there a market for jams? We know that lots of people like two-dollar pies and 42¢ scoops of ice cream.

452. If Jan had been saving her money over the years instead of buying "stuff," she might have been able to have enough to rent a small store space and get started.

If she had saved $130 per month and had earned 0.9% on her money each month, at the end of six years (72 months) she would have

$$130 \times \frac{(1 + 0.009)^{72} - 1}{0.009}$$

I'll do the hard part: $1.009^{72} \doteq 1.906$
Find out how much Jan would have at the end of six years.

Chapter Thirty-eight Better Bones

Second part: the 𝔐ixed 𝔅ag: a variety of problems from this chapter and previous chapters

532. In the previous problem, saving $130 each month and earning 0.9% on that money each month will give you at the end of 72 months

$$130 \times \frac{(1 + 0.009)^{72} - 1}{0.009}$$

This is called an annuity (if you want to look it up on the Internet). Now give me the **general formula for an annuity** if you save d dollars for n months and are earning i interest each month.
(In the previous problem, 0.9% was i = 0.009.)

602. Jan had received her last paycheck—$93.44. *At least I won't starve,* she thought to herself. How many two-dollar pies could she buy at Ivy's?

715. If an Ivy pie weighs 1.6 pounds, how much would a mole of those pies weigh?

880. It was 9 a.m. Jan's shift (and job) were over. She bought a Bison-Blueberry pie with a scoop of Vacant Vanilla on top. $2.42 (= $2 + 42¢)
   She sat at one of the outdoor tables at Ivy's and ate it. That was a meal for Jan. If she kept spending $2.42, and she started with $93.44, how many meals would she have?

Chapter Thirty-nine
Integumentary System

First part: Problems from this chapter

115. Jan washed some blueberry off her face. As she thought about how she might start a business named Jan's Jams, she ran her fingers through her hair. She looked at her nails. Skin, hair, nails—these are all a part of her _____.
fill in one word that begins with "i"

279. "Pardon me," he said. "I know that you must be in deep thought, but I was wondering if you might help me."

He smiled. Jan's mind went blank. Well . . . not exactly blank. Instead of thinking about her acting career, her former work at Ivy's Ice Cream, her poverty, her plans for Jan's Jams, there was one giant image that flashed into her consciousness.

He continued, "My mom is in the hospital and I was trying to think of some small gift that I might bring her when I visit her at noon. You are a woman. Do you have any thoughts?"

Thoughts! That word reverberated* in Jan's skull. (We won't say cranium, because that's more of a medical term.)

It's amazing how fast you can think when your mind is fully engaged. Jan's racing thoughts: I like his integument, especially his hair and his cheeks. He is kind—wanting to visit his mother. He called me a woman. He's talking to me. I hope he isn't married. He is polite; he said, "Pardon me." He recognized that I can have deep thoughts. I'm glad I don't have blueberry on my face.

Jan said the first thing that came to mind, "I would suggest some jam. That's always a good gift to bring to someone in the hospital."

Jan thought to herself: *I said that! He'll think I'm nuts.*

---

* *Reverberated* = with sound it means reechoed, with sound or light it means to be reflected many times, and with thoughts it means to have a lingering effect. With Jan his words were ringing like bells in her head—wedding bells.

Chapter Thirty-nine Integumentary System

He continued smiling and looking at her. "That's an interesting idea. Would you care to tell me more over a cup of hot chocolate? I hear that Stanthony's, which is just down the street, has great Italian hot chocolate."

Three seconds for her to stand up. Walking at 4 feet per second. Jan knew it would take a total of 51 seconds before they were at Stanthony's. How far would they walk?

368. They walked. He said, "Jam?"

She stopped and turned to look a him. (This walk will take more than a minute.) She has just said the first thing that had come into her head. Now she had to pretend that this was a well-thought-out response.

"Have you ever heard of Bison-Blueberry jam?" she asked. She wanted him to talk. That would take the pressure off of her.

"My sister, Ivy, has an ice cream store. You must know that store. You were sitting at one of her tables. She has just introduced Bison-Blueberry ice cream and Bison-Blueberry pie."

Jan was going to say, "~~Your sister just fired me~~," but mentally crossed that out.

Instead, "It's a new flavor of jam that I'll be introducing in my new business, *Jan's Jams*. My name is Jan."

"Wow. My name is also Jan. It's a form of John in some northern European countries."

They continued walking. He was thinking: *She's a businesswoman like my sister. She has great ideas and dreams. She's pretty.*

At Stanthony's it was Hot Chocolate Day: 25% off all hot chocolate drinks. They sat down at a table in the far corner of the restaurant. She ordered a Roma hot chocolate. Stanthony told her that was a good choice. (He told everyone that no matter what they ordered.) He said that after the discount it would only cost $1.95.

What was the price before the discount? (This is not a super easy question.)

439. On the menu the Roma hot chocolate had the description: *Make lots of room-a for the Roma!* She asked Stanthony what that meant. He told her that it comes in a large cup—1,200 milliliters. (This is sometimes written as 1,200 ml.) How many quarts is that? (1,000 ml = 1 liter. 1 liter = 1.06 quarts.)

Chapter Thirty-nine Integumentary System

Second part: the 𝔐ixed 𝔅ag: a variety of problems from this chapter and previous chapters

505. After Stanthony put that giant cup of Roma on the table, he turned to the male Jan and asked, "And what would you like?"

Male Jan said, "Two straws." Female Jan giggled. She liked the idea that they would be sharing. If they used their straws at the same time, his face would be close to hers.

He asked, "Tell me about your new business, Jan's Jams. I like that name."

She thought *If his sister is Ivy, a super successful business woman, I bet he knows a lot about the business world. I know zip.**

She began, "Right now we [meaning she] are in the first planning stages of Jan's Jams."**

"Great," he said. "It's always good to have a business plan." Ivy had told him about business plans, but he wasn't sure what they were.

As they talked, she drank one-seventh of the cup and he drank one-fifth of the cup. What fraction of the cup was left?

676. To the nearest percent, what fraction of the cup was left?

719. Neither of them really knew much about starting a business. Each thought the other one knew a lot. She switched the topic a little by saying that she had invented the Bison-Blueberry flavor and had given that idea to Ivy. She didn't mention that she had called Ivy at 1:20 in the morning to share that idea.

An hour went by. The *Roma* got cold. They didn't notice. At some point they were talking about Jans' Jams instead of Jan's Jams. That little change in the placement of the apostrophe meant a lot. *They* would be making the jams.

Jan offered his Disaster credit card to Stanthony. Stanthony thought to himself *This guy doesn't even carry two bucks with him.*

---

* *Zip* = zilch, which is American slang for zero or nothing.

** When you interrupt a quotation with words that are not in the original quotation, you use brackets, not parentheses.

Brackets ☞ [ ]   Parentheses ☞ ( )   Braces ☞ { }

Chapter Thirty-nine Integumentary System

It was true that he didn't have two dollars on him. All his money was back at his apartment. It was sitting in a pile on his dresser. It was a quarter and a penny.

Sometimes he put the quarter on top of the penny, and sometimes he put the penny on top of the quarter. He wondered which looked like more money.

She had $91.02.* He had $0.26. She had about 350 times as much money as he had. (91.02 ÷ 0.26 ≐ 350)

What percent more did she have than he had?

---

* Her paycheck was $93.44, and she had spent $2.42 on a Bison-Blueberry pie with a scoop of Vacant Vanilla on top.

Chapter Forty
Epidermis

First part: Problems from this chapter

164. $x^3 x^5 = ?$

240. "This might get a little confusing. We are both named Jan," she said.
　　He said, "It's even worse. My father's name is also Jan. He's Jan Senior and I'm Jan Junior."
　　"That makes things easier. I'll call you Junior and you can call me Janice, which is my regular name."
　　I, your reader, am also relieved. I was worried that for the rest of this book we would have male Jan and female Jan.
　　Junior said, "It's getting close to noon. I want to visit my mom in the hospital."
　　Janice had the perfect response: "May I come along?" Junior was very pleased.
　　Junior let Janice ride his bike. It was a mile (5,280 feet) to the hospital. He would jog. She got there in 440 seconds. That was 220 seconds before he did. How fast was each of them traveling? Give your answers in feet per second.

366. Janice was very happy that she was in shape. Years of dancing helped a lot. Junior was breathing pretty hard. They took the elevator rather than the stairs to the second floor of the hospital.
　　They walked down the hospital hallway. They passed a little boy with little purple ▭ marks all over his body. There were flowers on the floor.
　　Before they got to the room of Junior's mom, Janice asked him what his mom's name is. Junior blushed. "Her name is Janus. You can just call her mom." Junior has long eyelashes. His sister, Ivy, has short eyelashes. His mom, Janus, has short eyelashes. Two questions: Does Jan Senior have long eyelashes? What is the genotype of Junior?

Chapter Forty Epidermis

Second part: the 𝔐ixed 𝔅ag: a variety of problems from this chapter and previous chapters

425. Janus was in the hospital with a broken femur. What's a femur?

443. Janus gave her son a kiss. Junior said, "Mom, this is Janice."*
    After the introductions, Janus said to Janice, "I don't know whether my son told you how I got here. I was walking my dog and he saw a rabbit. He chased the rabbit and I was holding onto his leash. I busted my thighbone."**
    Janus turned to Junior and asked, "Have you been taking care of the dog since I got in here?"
    Attentive readers can now guess the name of her dog. Hint: The dog was born in the month of _____.

579. Let's take a break from all this and solve this equation:
    $5x + 0.6 + 7x = 39$

600. Balance this chemical equation in which carbon monoxide combines with hydrogen gas.
    $CO + H_2 \rightarrow CH_3OH$

---

* If you want to know how to do introductions right, here's the procedure. You first talk to the more important person and introduce the less important person to him/her.
    Then you introduce the less important person to the more important person.

** *Busted* is general (everyday) English. In general English you hear words like *ain't* and *busted* and *gonna*. In formal English you hear *are not* and *broken* and *going to*.

Chapter Forty-one
Dermis

First part: Problems from this chapter

138. Without looking back at the book, name the three things that the dermis produces that poke through the epidermis.

213. A bunch of tiny questions:
   A) Are skin cells made in the epidermis or in the dermis?
   B) Are there blood vessels in the epidermis or the dermis or both?
   C) We know that the dermis shoves three things *through* the epidermis. What two things does the dermis share with the epidermis?

336. Janus and everyone else were very happy that she didn't have a **compound fracture** of her femur. A compound fracture occurs when the sharp end of the broken bone breaks through the skin.

Officially ...

A compound fracture—also known as an open fracture—is any fracture in which there is a break in the skin near the broken bone. So a gunshot wound that breaks a bone is an open fracture.

The chance of **infection** *is the big concern with compound fractures for two reasons.*

#1: In a car accident, dirt, broken glass, grass, mud, and the person's clothing may be mixed into the wound.

#2: The sharp ends of the broken bone may tear up the nearby muscles, tendons, nerves, veins, and arteries.

Stir the germ-laden stuff of #1 into the mush of #2 and you can have a nightmare infection.

"Enough about my fracture and the dog," Janus said to her son. "Tell me about Janice." Moms can be very curious about their sons.

At first, Junior didn't know what to say. He didn't want to mention that they had just met hours ago. He hadn't been doing the **deep thinking** that Janice had been doing. He began, "Jan is a business women. She has great plans for a new business, which she'll call Jan's Jams."

Chapter Forty-one Dermis

Janice was relieved that Junior didn't know that she was just an unemployed ice cream scooper who had dreams about being an actress in the movies. That wouldn't have made a good impression on whom she hoped would be her future mother-in-law. (not "on who")

Junior looked like he was running out of things to say, so Janice took over their end of the conversation. She explained, "We are in the early planning stages of *Jan's Jams*. One main area of concern is finding money."

Janus smiled. She, like Ivy, knew a lot about running businesses. She said, "Oh, you're talking about venture capital. It's nice to know that you have all the other parts of a business plan in place."

Your question: Besides getting money to start a business, what are the other parts of a business plan? Translation: Besides getting some money to start a business, what other things need to be considered? Let's see how well your parents have been teaching you about business. I'm going to suggest seven other parts of any business plan. If you can name four, you are doing very well.

440. Janice didn't know what a business plan was. She thought that later she would ask Junior. She didn't realize that when it came to the business world, both of them were like three-year-olds in a calculus class. Is that a simile or a metaphor?

507. Janus was on pain medication. Breaking your femur is not fun. (litotes) She was starting to get sleepy. Janice and Junior had arrived there at 11:58 a.m. and it was now 12:32 p.m. It was time to leave. How long had they been visiting mom?

Chapter Forty-one Dermis

Second part: the 𝔐ixed 𝔅ag: a variety of problems from this chapter and previous chapters

607. They said goodbye and headed down the hallway, down the elevator, and outside. The bike wasn't there. Both had been so excited about their new "relationship" that neither of them remembered to lock the bike.

Some clown with *Gone with the Breeze* movie contracts in his pocket was now peddling around on Junior's bike.

Junior had owned a four-dollar bike and 26¢ in change. He now had 26¢. To the nearest percent, what percent of his "wealth" had he lost?

678. Junior wasn't unhappy. He wasn't very knowledgeable about running a business, but he did know a lot more about being happy than most people. He knew that: *It isn't the things that happen to you that determine whether you will be happy. It is how you respond that matters.*

There's an old saying that if someone steals your coat, give him your shirt also. Given the right state of mind, some clown stealing almost everything you own can't steal your joy.

The algebra is: Being in pain ≠ Being unhappy.

(≠ means "not equal")

Junior and Janice walked down the street. They were almost holding hands.

They were about to follow the Relationship Rule.

𝒯𝒽𝑒 𝑅𝑒𝓁𝒶𝓉𝒾𝑜𝓃𝓈𝒽𝒾𝓅 𝑅𝓊𝓁𝑒

The rule is: Not too soon and not too late.

When you first meet someone, you don't hang out all your dirty laundry. (metaphor) You don't tell them all the bad stuff about yourself. You don't tell them that you used to weigh 250 pounds or that you never clean up your room or that you have false teeth.

On the other hand, you don't wait till the wedding day to tell them that you are already married to someone else or that you never want to have children or that you are an alcoholic.

(continued on next page)

Chapter Forty-one Dermis

I, your author, hate numbers (irony alert), but if you need a number, then let's say that on the third time you meet, you should reveal the things that the other person should know.

Your question: Is "third" a cardinal number?

721. Janice and Junior had meet outside of Ivy's Ice Cream. They had spent an hour at Stanthony's enjoying a Roma hot chocolate. They had visited his mom in the hospital. They were now on a walk with each other. Without doing any counting—1, 2, 3—they began to follow the *Relationship Rule*.

He: You know, I really don't know that much about starting or running a business. My sister Ivy is the one who knows much more about those things.

She: I've got to confess that I also don't have the foggiest idea about such things. I just was tired of being poor and starting Jan's Jams seemed like a possibility. I'm glad that you have enough money to take me out to Stanthony's.

He: Money? I'm not rich. (litotes) In fact, I'm almost dead broke. My Disaster credit card is almost maxed out.*

They both felt better now. They each had less to hide. Neither had to pretend they were rich or knowledgeable about starting a jam business. One important part of any significant relationship—Trust—was forming.

Both had stuff they had to do this afternoon. Junior had to walk the dog January. Janice had to head back to her apartment. She wanted to see if there was anything that she might sell in order to raise some money.

She walked east at 4 feet per second. He walked west at 3 feet per second. In t seconds they were 4,620 feet apart. How long were they walking?

873. How many minutes did they walk? Use a conversion factor.

---

* Maxed out credit card = a card that has reached its credit limit. The credit limit on a card is the maximum amount that you can charge on the card. If the credit limit is $5,000, then you can't buy a $5,009 diamond ring with that card.

Chapter Forty-two
Genes

First part: Problems from this chapter

176. Junior got to his apartment and was greeted by January. January was with him while his mom was in the hospital.* They walked over to his dad's house to make sure he was okay.

    Senior was sitting very close to the television, but that's okay. His eyes and ears don't work that well.

    Senior watches everything that is on television. Junior only watches movies made in the 1940s. If A is the set of things Senior watches and if B is the set of things Junior watches, are those sets disjoint?

242. (continuing the previous problem) $A \cup B = ?$
(A ∪ B means the union of sets A and B.)

300. Suppose that there is a gene for dogs that gives them a curly tail. If T is the dominant gene that creates a curly tail and if t is the recessive gene, then since January has a curly tail, what must his genotype be?

449. If you were to take a heart cell from January and analyze the set of genes in that cell, could you tell whether January was TT or Tt?

583. Janice and Junior had agreed to meet at 5 p.m. at the park. There was so much they wanted to talk about.

    At her apartment Janice knew she had about four hours. She saw the place with new eyes. Half of her stuff was worthless and needed to be tossed out. A third of her stuff needed to be given to a thrift store. A tenth of her stuff was borrowed and needed to be returned to friends who had lent it to her. How much would she have left?

---

* Jan Senior couldn't take care of the dog on his own. He had a war-related head injury. All he could do is watch television and play video games. He was like some 13-year-olds. The only difference is that he had an excuse.

Chapter Forty-two Genes

Second part: the 𝔐ixed 𝔅ag: a variety of problems from this chapter and previous chapters

648. Janice's place had looked like this. ⇒

In a couple of hours it looked like this.

     She had discovered things that she didn't know that she owned. She had room to dance.
     Living in most American houses could be greatly improved by getting rid of 75% of the stuffing. Rooms would look bigger. There would be room to breathe.

If you haven't used it in a year,
seriously consider getting rid of it.

     Your question: Can you think of any exceptions to that rule?

714. B_2H_6 is a rocket fuel called diborane. The skeleton equation for burning diborane is $B_2H_6 + O_2 \rightarrow B_2O_3 + H_2O$ Balance this equation.

939. $y^3y^7 = ?$

955. Janice hopped into the shower. She uses one twenty-fourth of a bar of soap for every 5 minutes she's showering. She thought of meeting Junior at the park at 5 p.m. She dreamed. She sang. She spent an hour in the shower. How much soap did she use? (Use a conversion factor.)

962. Junior took a nap from 12:55 to 4:30. How long did he sleep?

Chapter Forty-three
Six Words

First part: Problems from this chapter

124. (Test to see how carefully you read the biology in Chapter 43) Everyone (now) knows that we have 23 pairs of chromosomes. We have an even number of chromosomes. Forty-six is an even number.*

Not all living things have an even number of chromosomes. The chapter gave two examples—living things with an odd number. Name them.

260. For thirty-five year-old mothers, $\frac{1}{900}$ of the pregnancies result in a baby with Down syndrome.

For forty-five year-old mothers, $\frac{1}{35}$ of the pregnancies result in a baby with Down syndrome.

To compare these fractions we *pretend* we are going to add them. We make their denominators alike.

$$\frac{1}{900} = \frac{35}{31500} \quad \text{We multiplied top and bottom by 35.}$$

$$\frac{1}{35} = \frac{900}{31500} \quad \text{We multiplied top and bottom by 900.}$$

This means that in every 31,500 conceptions, on the average, there will be 35 DS babies for thirty-five-old moms and 900 for forty-five-year-old moms.

Nine hundred is what percent greater than 35? Round your answer to the nearest percent.

---

* English note: Depending on whom you read, you usually write numbers greater than nine or ten as digits. I have 23 chromosomes. The exception is when you begin a sentence with a number. In that case you write it out. Forty-six is an even number. English always seems to have exceptions—except in a few cases.

Chapter Forty-three Six Words

Second part: the 𝔐ixed 𝔅ag: a variety of problems from this chapter and previous chapters

359. At 5 p.m. clean Janice was at the park. (She used a half of a bar of soap; you have to call her clean.) He wasn't there.
    She checked the playground.
        She checked the park bench.
            She checked the vending machine area.
        She started to cry.

    She waited. She thought *I really don't know that much about him. We've only known each other for hours. Maybe he's always late. Maybe he has forgotten about me. Maybe. . . .* (This is one reason you don't marry someone until you really know a lot about them. Knowing a lot about someone takes more than a couple of months.)

    Junior had been napping with a smile on his face. He was looking forward to seeing Janice whom he liked. He had set his alarm for 4:30, which gave him plenty of time to get to the park by 5. The phone rang. It was from the hospital. They said that his mom, Janus, had had a stroke.* He ran to the hospital (no bike). There was no way he could contact Janice. He didn't have her cell phone number.
    The doctor explained to Junior that his mom's stroke was not related to her broken leg. It just happened. It was lucky that it happened in the hospital. Janus could receive immediate medical attention.
    Junior was not used to doing much exercise. He was pooped. He had run 1.2 miles to the hospital at 8 mph (miles per hour). How many hours did it take him? How many minutes was that?

446. Janice had cried off her Ralph's fake eyelashes. She had never told Junior that she had short eyelashes, but he had already guessed. One of her fake eyelashes had been on crooked when they met.
    She didn't know where he lived or his phone number. She headed to Ivy's Ice Cream. Ivy should know her brother's phone number.
    Ivy wasn't there.
    Make a guess. Why wasn't she there?

---

* Stroke = either a blood clot in the brain or a broken blood vessel in the brain. Some strokes are mild and some are serious.

Chapter Forty-three Six Words

508. Janice asked one of the robots that* was running the store, "Where's Ivy?"

It answered, `"Ivy to the hospital went. Her brother's mom sick bad."`

Janice understood. She ran to the hospital. She wasn't pooped when she got there because she was in great shape. Her years of dancing had given her a strong heart, lungs and muscles.

There are two arteries coming out of her heart. Name them.

690. Janice didn't remember the room number for Janus. The nurse behind the counter was on the phone, talking, talking, talking. Janice didn't want to wait.

She spotted a nurse walking down the hall and asked her. The nurse said, "Yes. I'm heading there right now. Come with me."

The nurse took her to Jay Ness's room. Wrong room. Jay Ness wasn't Janus. Janice asked Jay if he was related to the famous Eliot Ness.** He shook his head.

Janice saw Ivy in the hallway. Janice ran to her. (Healthy adults and most kids love to run.) Ivy explained to Janice what had happened. Janice was very relieved that Junior had not just forgotten her. As they walked down the hall, they saw a sign: Ear plugs—.10¢ a pair. How many pairs could you buy for a dollar? (Hint: The answer is not ten.)

---

* If there were a human running the store, I would have written, "Janice asked one of the people *who*. . . ."

    Machines are *that*.
    People are *who*.

** Eliot Ness was a famous crime fighter in the 1930s. He fought against the crook Al Capone.

Capone didn't want to be captured by Ness. He offered Ness a deal: I'll put $2,000 on your desk every Monday if you stop chasing me. $2,000 in the 1930s is the same as $29,000 today. That's $29,000 *each week*. Ness said no.

Ness and his team of police were called The Untouchables.

Chapter Forty-four
Five Words

First part: Problems from this chapter

166. In the previous chapter (Chapter 43—*Six Words*) chromosomes were talked about. Each of us has 23 pairs. In this chapter (Chapter 44—*Five Words*) there were hopes that the other five words would be explained.

    locus
      site
        DNA
          gene
            allele

When Fred got all excited about his 23rd chromosome, none of those other five words were discussed.

At the end of Chapter 43 we had one word that was discussed.

At the end of Chapter 44 we still were discussing that one word. What percent increase in the number of words happened?

209. If *gene* had been introduced in Chapter 44, we would have increased the number of biology words from 1 to 2. (from just *chromosome* to both *chromosome* and *gene*) What percent increase would that have been?

337. Describe the 23rd chromosomes of Janice, Junior, Senior, and Janus.

490. The ability to see color is a dominant gene on the X chromosome. The Y chromosome doesn't have that gene at all. Let's call that dominant gene C (for color). The recessive gene we'll call c.

If you don't have C in your genotype, red and green are just shades of gray for you. You are color-blind.

Suppose that one-hundredth (1%) of all the X chromosomes in the world have c.

Then for a woman to be color-blind, both of her XX chromosomes would have to have c. The chances of that happening are $\frac{1}{100} \times \frac{1}{100}$ which is $\frac{1}{10,000}$ Rephrase the previous sentence in percents.

Chapter Forty-four Five Words

Second part: the 𝕄ixed 𝔹ag: a variety of problems from this chapter and previous chapters

584. (continuing the previous problem) What percent of men would be color-blind?

611. Ivy took Janice to Janus's room. Junior was there. Everyone was happy to see everyone else.

Janice thought *It would be wonderful to be a part of this family.*

The stoke had been mild. Janus was more interested in talking with her son and Janice about their dreams than about her medical condition. She said, "Junior has told me that neither of you know a whole lot about business and neither of you have a lot of start-up capital.* I'm sure that Janet will be happy to help you in both of these areas."

𝕁𝕒𝕟𝕖𝕥! I, your reader am going nuts. Are you, Mr. Author, going to introduce another Jan into the story this late in the book? You already have Janice and Jan Junior and Jan Senior and Janus and January and Jay Ness. Please be merciful. I don't want Janet to walk in the door.

No sweat. I don't think I could juggle another Jan either.

Then who is Janet?

Just a little patience. I'm going to put some flowers here and we'll continue the story.

❀   ❀   ❀

Janice almost screamed, "Who is Janet!!!!"

Janus said, "She's the one who owns the ice cream store. You've met her I'm sure."

Ivy explained to Janice, "When I was born, they named me Janet Ivy. I liked my middle name more than my first name, so I go by Ivy rather than Janet. My mother still likes to call me Janet. And, yes, I'd be glad to help you and my brother start a business."

There are two main reasons why most new businesses fail. One is the lack of enough start-up capital and the second is lack of knowledge about the business you are starting. Both of these are discussed in detail in *Life of Fred: Financial Choices*. ← real book. $19 (cheap!), 176 pages.

In that book it says, "If your field of interest is bowling alleys, you don't start by going out and buying a bowling alley." (p. 116)

Your question: How long, in weeks, is 5,000 hours of preparation?

---

* *Start-up capital* = money to start a business.

Chapter Forty-five
Five Words: Locus, Site, DNA, Gene, Allele

First part: Problems from this chapter

112. Each of the 46 chromosomes is just a big long DNA molecule. And each DNA molecule is a string of nucleotides (AGTCCCCATTCC...). Are all of the 46 DNA molecules the same length?

177. Five thousand hours of preparation = a decent amount of education needed to have a reasonable chance of success in starting a business.
    In a previous problem we computed that spending 40 hours per week getting educated about creating a *Jans' Jam* business would take about 125 weeks.
    Janice and Junior looked at each other. They both had the same idea: *What if we worked together? 62½ weeks. One of us could study advertising. One of us could study all the aspects of making jam. One, whether there is a market for jams—what kinds of jam, what prices might be charged. One, the available store spaces to rent. One, the government regulations on a jam business. One, whether a store or an Internet site would be preferable—or both. One, the accounting—what would be the expenses, what would be the expected income, what would be the profit. One, find out what is the current competition in the jam business—would people just prefer to buy their jam at the grocery store or spend more on a gourmet jam.*
    Your question: List the advantages and disadvantages of Janice and Junior doing all the stuff in the previous paragraph.

214. There are two things that Janice and Junior would find out if they chose to spend those 5,000 hours. First, they would find out whether *Jans' Jams* was a good business idea. Second, each would learn whether the other person was a good worker or just a lazy bum. Running a business takes a lot of work. Ivy has cut her hours down to "just" 12 hours/day, but she will retire rich. →→→ Most people who work for wages, do not retire rich. ←←←
    The advantages of retiring rich are obvious. List the disadvantages of retiring rich.

Chapter Forty-five   Five Words: Locus, Site, DNA, Gene, Allele

Second part: the 𝔐ixed 𝔅ag: a variety of problems from this chapter and previous chapters

360. Science is zooming along today faster than any previous time in all of human history. It was only very recently that a procedure was developed to manipulate specific genes.

Three chapters ago we learned about point mutations such as the one that causes cystic fibrosis.

In order to "tell" the procedure to work on that specific gene, you have to tell it which of the 23 pairs of chromosomes, and you have to tell it where on that chromosome to work. The location of a gene on a chromosome is called the _____ of that gene.
fill in one five-letter word

475. Janus was getting tired. A broken leg and a stroke can do that. Janice, Junior, and Ivy knew it was time to leave and let Mom get some rest.

nearby lake

Ivy headed back to her ice cream store. Janice and Junior took a walk around a nearby lake. They had a lot to talk about.

Janice said, "I've always wanted to have a number of children that was evenly divisible by 3."

Junior, "I want an even number of kids."

What are the first three possible numbers of kids that they would both be happy with?

582. They looked at a map of the lake. The map said that 3.4 cm (centimeters) corresponds to 5 miles. With a ruler they measured the distance around the lake. It was about 4 cm. To the nearest tenth of a mile, how far is it around the lake?

646. Junior had one thing on his mind. "Assuming for a moment that we might get married, would you be willing to take my last name?"

Janice thought *I really don't care what his last name is. I'd be happy to do that.*

She said, "Just out of curiosity, what is your last name? We've never really been formally introduced."

"It's Janson."

Chapter Forty-five   Five Words: Locus, Site, DNA, Gene, Allele

"That's a lovely name. Then my name would be Janice Janson. I really like that."

If it was 5.9 miles around the lake and if the lake were circular, how far is it across the lake? Use 3 for π in this problem. Round your answer to the nearest tenth of a mile.

980. They sat on a bench and looked over the lake. They discussed seriously things that might get in the way of their creating Janson's Jams. (That's much easier than saying Jans' Jams.)

They talked about everything that they could think of. They even talked about whether they squeezed the toothpaste tube in the middle or at the end.

They talked about their hobbies. He liked to sing in the church choir. (Janice once had a boyfriend who liked to spend all summer hunting elk. That wouldn't be good if they were going to be working hard to prepare to Janson's Jams.) He also liked watching movies from the 1940s.

"Why?" she asked.

Junior said, "Some of the finest musicals were in that era. The actors and actresses really knew how to sing and how to dance."

Janice smiled. She revealed that one of the big interests in her life was singing and dancing. "For a couple of weeks I sang and danced at your sister's ice cream place."

Then Junior said something that made Janice's heart rate go from 80 to 180. What percent increase is that?

❀   ❀   ❀

She asked him what he did for a living. He said that he had always liked singing and dancing. He was the chief talent scout for Major Music Movies. MMM was always looking for actresses who could really sing and dance.

And they lived happily ever after.

Complete Solutions and Answers | 106–115

106. A quart of litter comes in a 3-pound box. There was a hole in the box and 5 ounces had leaked out. How much was left?

```
  3 pounds   0 oz.         2 pounds  16 oz.
−          5 oz.        −           5 oz.
                           2 pounds  11 oz.
```

There would be 2 pounds, 11 ounces left.

110. What is the first thing that should be written down?

   The question asked was how long was it before Jan caught the movie producer.
   So the first thing to be written is:
   Let t = the number of seconds that they ran.

112. Each of the 46 chromosomes is just a big long DNA molecule. And each DNA molecule is a string of nucleotides (AGTCCCCATTCC. . .). Are all of the 46 DNA molecules the same length?

   In the previous chapter Fred found out that his Y chromosome was a "little runty" chromosome compared with his X chromosome.

114. $\frac{5}{6} + \frac{3}{4}$

$= \frac{10}{12} + \frac{9}{12}$ We multiplied the top and bottom of the first fraction by 2.
  We multiplied the top and bottom of the second fraction by 3.

$= \frac{19}{12} = 1\frac{7}{12}$

115. Jan washed some blueberry off her face. As she thought about how she might start a business named *Jan's Jams*, she ran her fingers through her hair. She looked at her nails. Skin, hair, nails—these are all a part of her integument .
fill in one word that begins with "i"

115

| 119-121 | Complete Solutions and Answers |

119. Find the LCD for each of these:

A) $\frac{5}{6}$ and $\frac{3}{8}$ The LCD is 24.

B) $\frac{3}{10}$ and $\frac{1}{4}$ The LCD is 20.

C) $\frac{7}{10}$ and $\frac{2}{15}$ The LCD is 30.

Different people find LCDs in different ways.

One way to find the smallest number that divides evenly into 6 and 8 is to list the multiples of 6, which are 6, 12, 18, 24, 30, 36, etc. and pick the first one out of that list that 8 divides evenly into.

121. If Ivy's were only open 23 hours each day, what percent of the time would she be open. Round your answer to the nearest percent.

23 is what percent of 24?
23 = ?% of 24

We have done the old "If you don't know both sides of the *of*, then you divide the number closest to the *of* into the other side" so much, it's time to look at it a little differently. We'll take a more algebraic approach.

Let x = the percent of the time Ivy's will be open each day.
Then 24x = the hours she will be open.*

The equation becomes   23 = 24x

Divide both sides by 24 $\frac{23}{24}$ = x

$\frac{23}{24}$ ≈ 0.958333 ≐ 0.96 = 96% of the time Ivy's will be open

---

* This line may need a little more explanation. We know that x is the percent of the time that Ivy's will be open. Say, for example, x were 30%. So 30% of the 24 hours her store would be open. Thirty percent of 24 is (0.3)24. Now replace the 30% by x. Her store will be open x(24), which can be written as 24x.

Complete Solutions and Answers | 122–124

122. There's a 0.2% chance Cassie won't be able to perform. That is one performance out of how many?

   One is 0.2% of how many?
   1 = 0.2% of ?

We don't know both sides of the *of*, so we divide the number closest to the *of* into the other number.

   $1 \div 0.2\%$

   $1 \div 0.002 = 500$

One out of every 500 performances Cassie won't be able to perform.

   This is an average. This doesn't mean that every 500th performance she will fail to perform. She might go 1200 performances and then have several times where she couldn't make it.

   They have done studies that show that on the average when you run a red light, there is one chance in 187 that there will be an accident in which there is a fatality. Thud (the guy who copied my middle school essay) once said, "I'm safe. I have only run 150 red lights in my life."

123. Jan bought 5 new shirts and a $6 hair ribbon for her new dancing job at Ivy's Ice Cream. The cost was $73. How much was each shirt? (The shirts were all the same price.)

   Let x = the cost of one shirt.
   Then 5x = the cost of 5 shirts.
   The 5x + 6 = the total cost.

The equation is now easy to write.   5x + 6 = 73
Subtract 6 from both sides   5x = 67
Divide both sides by 5   x = 13.40

   The cost of one shirt is $13.40.

124. Name two things with an odd number of chromosomes.

   On page 212 we learned that some plants are triploid—their chromosomes come in triplets. If some triploid plants have an *even* number of triplets, then they would have an even number of chromosomes since Even × Odd = Even. But if they have an odd number of triplets, then they would have an odd number of chromosomes since Odd × Odd = Odd.

   The second example was given on page 214: Down syndrome babies with 47 chromosomes instead of 46.

127–129 Complete Solutions and Answers

127. Midsummer Night's Cream. One gallon of cream. Fold in 5 ounces of whipped cream.

Cream comes in 10-liter containers. How many gallons is that? (Round to the nearest tenth of a gallon. One gallon is approximately 3.785 liters.)

The conversion factor will be either $\dfrac{1 \text{ gallon}}{3.785 \text{ liters}}$ or $\dfrac{3.785 \text{ liters}}{1 \text{ gallon}}$

$\dfrac{10 \text{ liters}}{1} \times \dfrac{1 \text{ gallon}}{3.785 \text{ liters}} \approx 2.642$ gallons $\doteq 2.6$ gallons.

128. One-third of the way through the meal she had already eaten 2,877 Calories. At that rate, how many Calories would she have for lunch?

First approach: You would have to triple the 2,877 to find the total Calories eaten. 3 × 2,877 equals 8,631 Calories. (This is the first *Life of Fred* book in which you are allowed to use your calculator.)

Second approach: You ask, "2,877 is one-third of ?"

$2{,}877 = \dfrac{1}{3}$ of ?

You don't know both sides of the *of*, so you divide the number closest to the *of* into the other number.

$2{,}877 \div \dfrac{1}{3} \quad \Rightarrow \quad \dfrac{2877}{1} \times \dfrac{3}{1} \quad \Rightarrow \quad 8{,}631$

I, your reader, think this second approach is worthless. It's a lot more work than just tripling the number. Why did you include it?

That's easy. One of the following is true:

A) I am a big, bad, mean author who just likes to inflict extra "stuff" on his readers OR
B) I knew what was coming in the next problem and wanted to make your road a little easier.

129. Ivy's Ice Cream offers 1,000 flavors. 57% of them are not very popular. How many flavors are not very popular?

What is 57% of 1,000?
0.57 × 1,000   If you know both sides of the "of," you multiply.
= 570 flavors are not very popular.

Complete Solutions and Answers | 130-133

130. You start with 10 pins. You knock over one pin. What percent of the pins remain standing?

Nine pins (out of 10) remain standing.
9 is what percent of 10?
9 = ?% of 10

Since you don't know both sides of the *of* you divide the number closest to the *of* into the other number.

$$9 \div 10 = \frac{9}{10} = 0.9 = 90\%$$

131. Is digesting a slice of pizza a voluntary or involuntary action?

Do you have to tell your stomach to work? Do you have to tell it when to send it to the small intestines? Do you have to tell the small intestines to share the good stuff in the food with your blood stream?

No. You can do digesting without having to think about it. It is an involuntary action.

132. If the carbon dioxide concentration were 0.1%, then plants could find a CO_2 in one out of every how many?

$$0.1\% = 0.001 = \frac{1}{1000}$$

One out of every 1,000 air molecules would be CO_2. Plants would be very happy.

Happy plants (like grass) → Happy cows → More ice cream

133. You arrived half way through the second act (of five acts). What percent of the play had you missed?

You had missed all of act one and half of the second act.
You had missed 1½ acts out of 5.
1½ is what percent of 5?
1½ = ?% of 5
1½ ÷ 5

If you used a calculator, $1.5 \div 5 = 0.3$

If you used a pencil, $1\frac{1}{2} \div 5 = \frac{3}{2} \div \frac{5}{1} = \frac{3}{2} \times \frac{1}{5} = \frac{3}{10} = 0.3$

In either case, 1½ ÷ 5 equals 30%.

| 134–138 | Complete Solutions and Answers

134. One neon atom has an atomic weight of 20.18. What is the atomic weight of a dozen neon atoms?

$12 \times 20.18 = 242.16$ is the atomic weight

135. Reduce this fraction $\dfrac{31688113005897}{31688113005897}$

There are four things to remember about fractions:

- Reduce fractions in your answers as much as possible.
- Fractions like $\dfrac{0}{4}$ are equal to 0.
- Fractions like $\dfrac{4}{4}$ are equal to 1.
- Division by zero is not permitted.

Using the third thought $\dfrac{31688113005897}{31688113005897}$ reduces to 1.

136. Newborns have about 350 bones. Adults have 206. What percentage of bones are "lost." Give your answer to the nearest percent.

144 bones are "lost." $(350 - 206 = 144)$
144 is what percent of 350?
$144 = ?\%$ of 350
$144 \div 350 \approx 0.4114285 = 41.14285\% \doteq 41\%$

137. Dimensional analysis and unit analysis are fancy names for what we have been doing for a long time. It is called using a conversion factor.

138. Without looking back at the book, name the three things that the dermis produces that poke through the epidermis.

Hair, sweat, and oil.

Complete Solutions and Answers | 150–152

150. Suppose I divide the entire physical universe into these two sets: Set A = everything that weighs less than 4 kilograms and Set B = everything that weighs more than 3 kilograms.

    Is the union of A and B equal to the entire physical universe?

    Yes. Every item in the physical universe must be in either set A or set B (or in both).

    Are A and B disjoint?

    No. There are items that are in both sets. For example, my breakfast pizza that I had this morning weighed 3.4 kilograms. It is in both sets.

Discussion: A kilogram weighs a little more than two pounds.
When you draw a Venn diagram of sets C and D and they are disjoint, it will look like this:

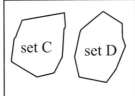

The sets C and D do not overlap. They have nothing in common.

151. Looking at wheat → mouse → cat → wheat → . . . we can think of it as a giant circle of life. The plants combine carbon dioxide and water and produce sugars, starches, and oils. Animals eat the sugars, starches, and oils and give off carbon dioxide. Like a wheel that it spinning, it would come to a stop unless something keeps it spinning. Where is energy inserted into this wheel of life?

    In photosynthesis, carbon dioxide and water and *energy from the sun* are converted into sugars, starches, and oils.

    It is the sun that spins the wheel of life.

152. You phoned her at 10 a.m. the next morning and woke her up. Jan had gone to sleep at 11:17 p.m. How long had she slept?

    From 11:17 p.m. to midnight is 43 minutes.
    From midnight to 10 a.m. is 10 hours.
She had slept 10 hours and 43 minutes.

153–154 Complete Solutions and Answers

153. $34 per month for 1,044 months." How many years is that?

We will solve this two different ways.

Way #1: Using the General Rule,* suppose it was 36 months. That would be 3 years. We divided by 12.

$1,044 \div 12 = 87$. It would take Jan 87 years to buy that ring.

Way #2: Using a conversion factor:

$$\frac{1044 \text{ months}}{1} \times \frac{1 \text{ year}}{12 \text{ months}} = 87 \text{ years}$$

Assuming Jan doesn't go bear hunting and lives another 60 years, we know that her children will be paying for that ring for another 27 years after Jan has died. $(87 - 60 = 27)$

154. Cassie had short eyelashes. Her parents both had long eyelashes. What is the probability that Cassie's brother had short eyelashes?

Cassie's genotype must be SS. (If it were either SL or LL, then she would have had to have long eyelashes since L is dominant.)

Each of her parents must have had an S (otherwise, Cassie couldn't have had SS.)

Each of her parents must have at least one L since they both have long eyelashes.

Therefore, each of her parents is SL.

short eyelashes

When two people with SL mate, there are four possibilities:

S from dad, S from mom	➭ kid has SS and short eyelashes
S from dad, L from mom	➭ kid has SL and long eyelashes
L from dad, S from mom	➭ kid has LS and long eyelashes
L from dad, L from mom	➭ kid has LL and long eyelashes

So there is one chance in four (25%) that Cassie's brother will have short eyelashes.

---

* If you don't know whether to add, subtract, multiply or divide, first restate the problem with really simple numbers.

Complete Solutions and Answers | 155–157

155. Ivy now had a 12-hour work day (instead of a 24-hour work day). If she slept for 8 hours, what percent of each day would be free? (Translation: What percent of the day would she not be either working or sleeping?)

    Four hours of each day she would be free. $(24 - 12 - 8 = 4)$
    4 is what percent of 24?
    $4 = ?\%$ of 24
    $4 \div 24 = \frac{4}{24} = \frac{1}{6} = 16\frac{2}{3}\%$

    This is one of the eleven conversions that you might have memorized in *Life of Fred: Decimals and Percents*. If you didn't, you could always just do the division:

$$6 \overline{)1.00} \quad 0.16 \text{ R}4$$

$$\frac{1}{6} = 0.16\frac{4}{6} = 0.16\frac{2}{3} = 16\frac{2}{3}\%$$

156. What number(s) make $0x = 0$ true?

    Let's try some numbers.
        If $x = 4$, then $0(4) = 0$ is true.
        If $x = 17$, then $0(17) = 0$ is true.
        If $x = 0$, then $0(0) = 0$ is true.
    It seems that every number makes $0x = 0$ true.

157. You take a slurp of Ivy's Sunshine in Your Tummy ice cream. It starts in your mouth.
    Arrange the rest of these in the order that the ice cream visits them:
        4 small intestine
        1 pharynx
        3 tummy
        5 large intestine
        2 esophagus

158–159 Complete Solutions and Answers

158. How many terms in each of these expressions?

6y + 32 – 20 __3__

$5x^{23}yz^4$ __1__

5 + 7 __2__   If you were to add these together and get 12, then the number of terms would change.

(3x + 6)(444) __1__   This is harder than the previous questions.

Advanced lesson for determining how many terms:

Replace any + or – sign by a big black bar ▌ *as long as the resulting pieces make sense.*

For example, 6y + 32 – 20 would become 6y ▌32 ▌20

So this has three terms.

For example: $5x^{23}yz^4$ doesn't have any + or – signs, so you couldn't stick any big black bars. It has one term.

For example: (3x + 6)(444) you could *not* replace the + sign with a big black bar because the resulting pieces would not make sense.

If I try to replace the + sign, I get   (3x ▌6)(444).
(3x doesn't make sense. Nor does 6)(444).

159. A scoop of Ivy's Ice Cream cost 47¢.

In the first hour that Jan worked she sold one scoop to Joe, one scoop to Darlene, and scoops to other customers. She also sold a 17-cent polar bear sticker to a girl for her sticker collection. The polar bear reminded her of ice cream. These sales totaled $13.80. How many scoops did Jan sell?

Let x = the number of scoops Jan sold.
Then 47x = the cost of those scoops.
Then 47x + 17 = the sales total.

Now we can write the equation   47x + 17 = 1380
<small>Subtract 17 from both sides</small>   47x = 1363
<small>Divide both sides by 47</small>   x = 29

Jan sold 29 scoops.

Complete Solutions and Answers | 160–162

160. Why haven't we run out of oxygen?

    Back in Chapter 17 we learned that besides us (who do a lot of breathing) there are plants. We use up a lot of oxygen and produce a bunch of carbon dioxide. Plants use up a lot of carbon dioxide and produce oxygen.

    Without us breathers plants would be unhappy.
    Without those plants we would be—the technical term is *dead*.

    We say to plants, "Give us lots of oxygen."
    Plants say to us, "Give us lots of carbon dioxide."
    It's a very nice trade.

    One fun science fair project is to grow plants in an atmosphere with extra carbon dioxide in it. They love it.

161.

$$9w - 3 = 4w$$

Add 3 to both sides $\qquad 9w = 4w + 3$

Subtract 4w from both sides $\qquad 5w = 3$

Divide both sides by 5 $\qquad w = \dfrac{3}{5}$

162. Balance the skeleton equation $H_2O_2 \rightarrow H_2O + O_2$

The H's are in balance. There are two on each side.
There are 2 O's on the left and 3 O's on the right.
Let's add some O to the left side $\qquad 2H_2O_2 \rightarrow H_2O + O_2$

4 O's on the left and 3 O's on the right.
Add some O to the right side $\qquad 2H_2O_2 \rightarrow 2H_2O + O_2$

The H's are in balance (4 on each side). The O's are in balance (4 on each side). ☺

There are many ways to try to balance a chemical equation. Someone might have gone back and forth several times and the final equation might be $16H_2O_2 \rightarrow 16H_2O + 8O_2$. That answer is the same as mine if you then divide through by 8.

| 163 | Complete Solutions and Answers |

163. **3 pies and 8 scoops of ice cream for $9.36!**
    Jan knew that scoops of ice cream are 42¢. Showing your work and beginning with *Let x = the price of a pie*, find the cost of one Ivy pie.

One little note before we begin. With a little head scratching you could probably figure out that the pies were $2 each. The whole point is to learn to do this using algebra and equations. When we get to "real life" the numbers wouldn't be as easy as this problem. It's important you learn how to translate the English into algebra.

    Let x = the price of a pie
    Then $3x$ = the price of 3 pies
    Then 8 scoops of ice cream = 336¢   (8 × 42¢ = 336¢)
    Then $3x + 336$ = the total cost

Now we can write the equation $3x + 336 = 936$
    Subtract 8 from both sides $3x = 600$
    Divide both sides by 3 $x = 200$

A pie costs 200¢. (or $2)

    Time out! I, your reader, think that you, Mr. Author, are laying it on a little thick.* All these algebra problems look simple-dimple. They all look like "Five hamburgers plus a $6 milkshake cost $33.35. How much does one hamburger cost? I'm bored!

Is this all there is to algebra? ⇨

If it is, then I don't need it.

> Let x = the cost of one hamburger
> Then $5x$ = the cost of 5 hamburgers
> Then $5x + 6$ = the total cost
> The equation is $5x + 6 = 33.35$
> $5x = 27.35$
> $x = \$5.47$

    I have been trying to be gentle. This is only the second of the three pre-algebra books. Any algebra we do here is just a foretaste of what will be taught in detail in *Life of Fred: Beginning Algebra Expanded Edition*. Word problems are probably the hardest part of beginning algebra. My presenting word problems here in pre-algebra is being very bold.
    Bold? I told you I was bored. What are you going to do about that?

---

* Laying it on thick = hyperbolizing = overselling = embroidering = exaggerating

Complete Solutions and Answers — 163

Do about that? I could ask you to read faster?

That's a wimpy answer. Give me a word problem that has a little SPICE in it. A little ZING. A little challenge.

I'm afraid. Most of my readers are just plain happy doing 7 cupcakes + two dollars worth of Sluice cost $9.70.

Stop it! Anyone can do Let x = cost of cupcake; Then 7x = cost of 7 cupcakes; Then 7x + 2 = total cost. 7x + 2 = 9.70   7x = 7.70   x = $1.10. **I demand more!**

Let me make a deal with you. Let's hide this back here in the answer section of the book. I won't put anything like this in the questions section of the book. You don't tell anyone that I'm doing serious algebra in this pre-algebra book. Is that a deal?

It's a deal.

Just so that no one will accidentally turn to this section of the book and freak out, I'm going to put the famous warning border on the right side of this page.

That border means that only **A** students are allowed to read this material.

Okay. Enough. Get on with it.

Two hamburgers and six shakes cost $44.
Three hamburgers and four shakes cost $41.

How much do each of these cost?

Let x = the cost of a hamburger
Let y = the cost of a shake

$$\begin{cases} 2x + 6y = 44 \\ 3x + 4y = 41 \end{cases}$$

Wait a minute! That's two equations with two unknowns. We haven't had that yet.

Are you asking me to stop?

Gulp. Now I'm the one who is scared.

Do you see the warning border on the right side? You can stop reading anytime you like.

Multiply the first equation by 3.
Multiply the second equation by 2.

May I ask why?

127

163   Complete Solutions and Answers

No. It will become clear in a moment.

$$\begin{cases} 6x + 18y = 132 \\ 6x + 8y = 82 \end{cases}$$

Subtract the bottom equation from the top equation.
$$10y = 50$$

The rest is simple-dimple (as you call it).
$$y = 5$$

What about x?

Do you remember the original equations?
$$\begin{cases} 2x + 6y = 44 \\ 3x + 4y = 41 \end{cases}$$

Just stick y = 5 into the first equation.

$$\begin{aligned} 2x + 6(5) &= 44 \\ 2x + 30 &= 44 \\ 2x &= 14 \\ x &= 7 \end{aligned}$$

But . . . but . . . but what if you stick it in the second equation?

Big deal. You get the same answer.
$$3x + 4y = 41$$
becomes
$$\begin{aligned} 3x + 4(5) &= 41 \\ 3x + 20 &= 41 \\ 3x &= 21 \\ x &= 7 \end{aligned}$$

If you would care to do one . . .
   Four bikes and three smart TVs cost $3955.
   Three bikes and two smart TVs cost $2810.
How much does each cost?
You find the answer worked out in problem #989.

Complete Solutions and Answers | 164–177

164. $x^3 x^5 = (xxx)(xxxxx) = x^8$

166. At the end of Chapter 43 we had one word that was discussed.
At the end of Chapter 44 we still had just one word.
What percent increase in the number of words happened?

The change in the number of words was zero.
Zero is what percent of the six words?
$0 = ?\%$ of 6
$0 \div 6 = 0$ There was a zero percent increase in the number of words.

176. Senior watches everything that is on television. Junior only watches movies made in the 1940s. If A is the set of things Senior watches and if B is the set of things Junior watches, are those sets disjoint?

No. Since Senior watches everything, he would also watch movies made in the 1940s. Two sets are disjoint only if they have no members in common.

177. Janice and Junior looked at each other. They both had the same idea:
What if we worked together? 64½ weeks. One of us could study advertising. One of us could study all the aspects of making jam. One, whether there is a market for jams—what kinds of jam, what prices might be charged. One, the available store spaces to rent. One, the government regulations on a jam business. One, whether a store or an Internet site would be preferable—or both. One, the accounting—what would be the expenses, what would be the expected income, what would be the profit. One, find out what is the current competition in the jam business—would people just prefer to buy their jam at the grocery store or spend more on a gourmet jam.

Your question: List the advantages and disadvantages of Janice and Junior doing all the stuff in the previous paragraph.

Advantages:
1. The big advantage is that they would greatly increase their chances of success. Just borrowing $100,000 from Ivy and buying some business cards—and not knowing the difference between grape jam and granola—would almost certainly doom them to failure.
2. They would get a good chance to really get to know each other. They would see each other in both good times and stressful times—a good investment in marriage preparation.
3. It might turn out that Jans' Jams is not a good business opportunity. They wouldn't waste $100,000.

Disadvantages:
1. Forty hours a week for a little over a year is a lot of effort. If they both just got jobs, that's usually easier than studying eight hours/day and gives them some current income.

179 | Complete Solutions and Answers

179. The atomic weight of Na (sodium) is 23. The atomic weight of Cl (chlorine) is 35. How many grams does that one molecule of salt weigh?

The molecular weight of NaCl is 58. (23 + 35)
One mole of salt weighs 58 grams.
One mole of salt contains 6.0221236×10^{23} molecules of salt.

We want to convert one molecule of salt into grams.

First, we convert a molecule of salt into moles of salt.

$$\frac{1 \text{ molecule of salt}}{1} \times \frac{1 \text{ mole}}{6.0221236 \times 10^{23} \text{ molecules}}$$

$$= \frac{1}{6.0221236 \times 10^{23}} \text{ moles of salt}$$

Second, we convert moles of salt into grams.

$$\frac{1 \text{ mole of salt}}{6.0221236 \times 10^{23}} \times \frac{58 \text{ grams of salt}}{1 \text{ mole of salt}}$$

$$= \frac{58 \text{ grams of salt}}{6.0221236 \times 10^{23}}$$

$$\doteq \frac{9.63}{10^{23}} \text{ grams in one molecule of salt}$$

I won't write that out as 0.0000000000000000000000963, because that's too hard to read.

You lied! I, your reader, protest. You said you wouldn't write it out and you did.

> *Intermission*
> *A small lesson about lying.*
>
> To utter a lie, two things are necessary. First, the speaker must know that what is being said is not the truth. Second, the speaker must believe that the hearer will accept that untruth.
> When I tell you that my brains are made out of pickles, it's not true, but I don't think you will believe me. It's not a lie.

Actually, my brains are made out of pizza. I'm not lying.

Complete Solutions and Answers | 188–189

188. Jan's poor management of her money was **an albatross around her neck**. Is this a simile or a metaphor?

If we had said that it was *like* an albatross around her neck, it would be a simile.

Instead, we said that it *was* an albatross. This is a metaphor.

I, your reader, have got a better question. What in blazes does that "albatross around her neck" mean?

I'm glad you asked. An albatross is a seabird. Sometimes it will follow ships at sea.

That's nice, but that doesn't explain this around-the-neck business.

I'm not done. A little over two hundred years ago (1798) Samuel Taylor Coleridge wrote a poem *The Rime of the Ancient Mariner*. I read that poem in high school. I read it many years after he wrote it.

Of course! Please get to the point.

I'm sorry. I liked that poem.

In that poem one of the seamen (known as the ancient mariner) shoots an albatross with a crossbow. That was not cool. Albatrosses were supposed to be a sign of good luck.

After the albatross is killed, all kinds of bad things happen to the ship. The other sailors blame the bad luck on the ancient mariner. They hang the dead bird around his neck.

After that poem was written, "hanging an albatross around a neck" has been a metaphor for a burden that feels like a curse.

In today's world . . .

✓ Alcohol is an albatross around the neck of alcoholics.

✓ All the taxes, which take over 50% of the income of many productive members of our society, is an albatross around their necks.

✓ You do not want to be an albatross around your parents' necks.

189. A) If you put feathers in your hair, does that make you a bird? No.

B) If you steal things, does that make you a thief? Yes.

C) If you know that $\log 10^e = e$, does that make you look smarter? Sure it does. Few people know that.

D) If you are good at acting and singing, does that mean you will get a part in a movie? Many things in life don't turn out the way we want.

| 204–207 | Complete Solutions and Answers |

204. This last summer Jan was in a show. It went three weeks before it closed. Each night from 6 p.m. to 10 p.m. for 21 nights. "I made $25 each night," Jan told you. How much did Jan make in those three weeks?

In this simple problem, you could just multiply $25 times 21 performances and get $525.

When the problems get a bit more complicated, using a conversion factor (or conversion factors) might make things easier.

The conversion factor is either $\dfrac{\$25}{1 \text{ night}}$ or $\dfrac{1 \text{ night}}{\$25}$

We start with 21 nights can we want to convert that into dollars.

$\dfrac{21 \text{ nights}}{1} \times \dfrac{\$25}{1 \text{ night}}$ We choose the conversion factor so that the "nights" cancel.

$= \dfrac{21 \cancel{\text{nights}}}{1} \times \dfrac{\$25}{1 \cancel{\text{night}}}$

$= \$525$

205. The dress rehearsal went from 6 p.m. to 10:25 p.m. How long was it?

From 6 to 10 is 4 hours. From 10 to 10:25 is 25 minutes. The rehearsal was 4 hours and 25 minutes.

206. Solve $4x + 7 = 29.8$
Subtract 7 from both sides $4x = 22.8$
Divide both sides by 4 $x = 5.7$

$\begin{array}{r} 5.7 \\ 4\overline{)22.8} \\ \underline{20} \\ 28 \\ \underline{28} \end{array}$

You can either divide 22.8 by 4 the old-fashioned way or you can use a calculator.

207. Out of Jan's paycheck the government took 20% in taxes. How much was left for the three weeks' of work?

If 20% was taken, then 80% was left.
80% of $525
When you know both sides of the *of* you multiply.
80% × 525 = 0.80 × 525 = $420.

132

Complete Solutions and Answers 209–214

209. Increasing from 1 to 2 is what percent increase?

Going from 1 to 2 is an increase of one.
One is what percent of the original "1"?
1 = ?% of 1
$1 \div 1 = 1 = 100\%$ increase

210. Will a reduction have to be performed on Fred?

A reduction is the first medical step done for someone with a broken bone. A reduction is the putting back together of the broken ends of the bone. Since Fred didn't break any bones, he didn't need a reduction.

213. A bunch of tiny questions:
    A) Are skin cells made in the epidermis or in the dermis?
        They are made in the epidermis.

    B) Are there blood vessels in the epidermis or the dermis or both?
        Only the dermis has blood vessels.

    C) We know that the dermis shoves three things *through* the epidermis. What two things does the dermis share with the epidermis?
        Because the epidermis doesn't have blood vessels, it gets the food and oxygen that it needs from the dermis.

214. ➔➔➔ Most people who work for wages, do not retire rich. ⬅⬅⬅
The advantages of retiring rich are obvious. List the disadvantages of retiring rich.

Small Essay While I Try to Think of
the Disadvantages of Retiring Rich

Of course, if you want to retire with lots of money, you will have to get that money somehow. Marrying someone who is rich is one way. Robbing banks is a second way—although it's tough to spend the money if you're in prison. Inheriting a fortune is a third way, but that requires the luck of having rich ancestors. Being self-employed is the surest way for 99% of people.

Okay. Here's my list of the disadvantages of retiring rich:
    1.

217–220 Complete Solutions and Answers

217. Seventeen of the flavors are popular. (They are ordered more than 40 times per month.) What percent of the flavors are popular?

17 is what percent of 1,000?

17 = ?% of 1,000   If you don't know both sides of the "of," you divide the number closest to the "of" into the other number.

$17 \div 1{,}000 = 0.017$

$0.017 = 1.7\%$

218. $\dfrac{5}{6} \times \dfrac{3}{4}$

$= \dfrac{5}{\underset{2}{\cancel{6}}} \times \dfrac{\cancel{3}^{1}}{4} = \dfrac{5}{8}$

219. Lysander (played by Alex) has 476 lines to memorize. He could memorize 56 lines per hour. How long (in hours and minutes) will it take him to learn his part?

We start with 476 lines and want to convert that into time.

$$\dfrac{476 \text{ lines}}{1} \times \dfrac{1 \text{ hour}}{56 \text{ lines}}$$

$= \dfrac{476 \,\cancel{\text{lines}}}{1} \times \dfrac{1 \text{ hour}}{56 \,\cancel{\text{lines}}}$

$= 8.5$ hours

8.5 hours equals 8 hours plus 0.5 hours.

$$\dfrac{0.5 \text{ hours}}{1} \times \dfrac{60 \text{ minutes}}{1 \text{ hour}} = 30 \text{ minutes}$$

It will take Alex 8 hours and 30 minutes to learn his part.

220. A carton of 𝖺𝗇𝖼𝗁𝗈𝗏𝗒-𝗅𝖺𝗆𝖻 𝗐𝗂𝗍𝗁 𝖺 𝗋𝗂𝖻𝖻𝗈𝗇 𝗈𝖿 𝗒𝖺𝗆 ice cream measures 2.2" × 3" × 4.8". (" means inches)

What is the volume of this box?

The formula for the volume of a box is length times width times depth. $V_{box} = \ell w d$

$V_{box} = \ell w d$ becomes $V_{box} = 2.2 \times 3 \times 4.8$
$V_{box} = 31.68$ cubic inches

This can also be written as $V_{box} = 31.68$ in^3

Complete Solutions and Answers 221–222

221. Jan saw you come in and sit down. She waved to you. You are not supposed to do that when you are acting on stage. Her salary for the night was $87. They fined her $100 for her waving. Her gain for the evening was a negative number. What was that number?

A gain of $87 and a loss of $100, result in a net loss of $13.
Her gain was –13 dollars.

Losing 13 dollars can be considered a gain of –13 dollars.

$87 - 100 = -13$

When your elementary school teacher told you that you can't subtract seven from three, she was trying to protect your delicate young mind from reality. Anyone who takes algebra knows that $3 - 7 = -4$.

In beginning (first-year) algebra, your teacher will tell you that you can't take the square root of a negative number. $\sqrt{-4}$ doesn't exist. Your advanced (second-year) algebra teacher will tell you that your beginning algebra teacher was trying to protect your delicate young mind from reality. Everyone who has had all of high-school algebra knows that $\sqrt{-4}$ equals $2i$.

Right now I can't tell you what i means. I need to protect your delicate young mind.

222.
1. You heard the refrigerator.
2. You smelled the kitty liter boxes.
3. You knew that you were standing straight up.
4. You felt a fly land on your forehead.
5. You couldn't see a thing.

A. mechanoreceptor
B. proprioceptor
C. photoreceptor
D. audioreceptor
E. chemoreceptor

1D Audioreceptors are involved in hearing.
2E Chemoreceptors are involved with taste and smell.
3B Proprioceptors tell you what position your body parts are in.
4A Mechanoreceptors give you the sense of touch.
5C Photoreceptors can detect light.

Without proprioceptors you might walk around with your arm sticking straight out and not know it. This can be very embarrassing on a crowded bus.

135

| 223–224 | Complete Solutions and Answers

223. Name five other special projects that a person might have. Be creative.

Here is some that I thought of. Yours will probably be different.

★ Becoming a good cook (especially pizza)
★ Playing croquet well
★ Learning the history of your favorite country
★ Studying metamathematics
★ Gaining skills as a welder
★ Collecting buttons from around the world
★ Become an expert on merry-go-round construction
★ Composing music for movies
★ Investigating what it means to become a world-class matador or
                                          car salesman or
                                          buffalo hunter or
                                          ballet dancer or
                                          onion farmer

Each of these special projects can provide great happiness for certain individuals. The important thing is to find a special project that <u>you</u> yourself enjoy.

224. How much would the total cost be spending $34 per month for 1,044 months?

We will solve this two ways.

Way #1: Using the 𝔊𝔢𝔫𝔢𝔯𝔞𝔩 𝔑𝔲𝔩𝔢, how much would $2 per month for 5 months be? Ten dollars. We multiplied.
$34 × 1,044 months = $35,496.

Way #2: Using a conversion factor:

Since 1,044 months = the whole purchase, the conversion factor will be either $\dfrac{1044 \text{ months}}{\text{whole purchase}}$ or it will be $\dfrac{\text{whole purchase}}{1044 \text{ months}}$

$\dfrac{\$34}{1 \text{ month}} \times \dfrac{1044 \text{ months}}{\text{whole purchase}} = \$35{,}496$ for the whole purchase

Complete Solutions and Answers | 225–227

225. Carbon monoxide combined with oxygen will produce carbon dioxide. Balance the skeleton equation

$$CO + O_2 \rightarrow CO_2$$

The C's are in balance. There is one on each side.
There are 3 O's on the left side; 2 O's on the right.
Let's add more O's to the right side.

$$CO + O_2 \rightarrow 2CO_2$$

There is 1 C on the left; 2 C's on the right.
Let's add more C's on the left side.

$$2CO + O_2 \rightarrow 2CO_2$$

The C's are in balance (2 on each side). The O's are in balance (4 on each side). ☺

226. The blood that comes out of your heart and through your aorta is at high pressure—around 120. On the way back to your heart the blood travels through your veins at a lower pressure.

Now if you were designing a human body—let's say the blood vessels in an arm—would you put the arteries near the skin and the veins on the inside or the other way round?

If you get a small cut on your arm, the blood will be darker red and will flow slowly. That's easy to fix. In first-aid courses they teach you to apply pressure to that wound to stop or slow that bleeding. Small cuts on the arm will hit the veins, not the arteries.

A deep wound to your arm might hit an artery—bright red blood and it is gushing.

A well-designed body has the arteries well protected.

In first-aid courses you learn that you can . . .

- ☞ bleed to death in seconds
- ☞ suffocate (no air) to death in minutes
- ☞ die of thirst in days
- ☞ starve to death in weeks.

227. What is the weight (in grams) of a mole of neon?

A mole of neon weighs 20.18 grams.

230–231 Complete Solutions and Answers

230. Where do you think Jan will be next Monday?

After Jan signed, the man in the clown outfit explained to Jan that there would need to be an "arrangement fee" of $100 to cover preliminary expenses of signing. He told her that she would get that back when she started work next Monday.

He told her that he would come back in an hour with the airplane ticket.

Jan phoned Ivy and explained to her that she needed a hundred dollars right now. She asked permission to take it from the cash register and then have it deducted from her $193.44 pay that she would receive at the end of her shift. Ivy said yes.

Jan handed him the money, kissed him on the cheek, and said, "I'll see you in an hour."

He never came back.
Jan would be scooping ice cream at Ivy's next Monday.

Because someone says they are a movie producer, doesn't make him one. Because someone looks like a movie producer, doesn't make him one. Because someone can pull out a piece of paper that says it is a movie contract, doesn't make him a movie producer.

The rest of the story: Jan eventually realized what had happened. She cried. Ivy suggested that she call the police.

The police came and asked her to describe the bad guy. She said that he was dressed as a clown. That wasn't a very good description. But they knew about this crook. He would dress up in funny outfits and find women who were desperate to become movie stars. He had a pocket full of movie contracts that he had printing up at the local stationery store.*

He was making about $100 per hour doing this scam.

His name was C.C. Coalback.

231. How far did Jan run?

Our first line was: Let t = the number of seconds that they ran.
Our second line will be: Then 12t = the number of feet that Jan ran.
We know that distance equals rate times time. d = rt
If Jan's rate is 12 feet per second, then d = 12t.

---

* *Stationery* deals with envelopes. See the "*e*." *Stationary* means standing still. This is a mnemonic.

Complete Solutions and Answers | 234–237

234. There were 38 50-pound sacks of flour. How many tons is that? (1 ton = 2,000 pounds)

38 50-pound sacks = 1,900 pounds
We want to convert 1,900 pounds into tons.
The conversion factor will be either $\dfrac{1 \text{ ton}}{2,000 \text{ lbs.}}$ or $\dfrac{2,000 \text{ lbs.}}{1 \text{ ton}}$

$\dfrac{1,900 \text{ lbs.}}{1} \times \dfrac{1 \text{ ton}}{2,000 \text{ lbs.}} = 0.95$ tons (or $\dfrac{19}{20}$ tons)

235. A) Could she divide 19,005 ice cream cones into two even piles?
B) Could she divide 19,005 ice cream cones into three even piles?
C) Could she divide 19,005 ice cream cones into five even piles?

A) No. 19,005 is not an even number. It doesn't end in 0, 2, 4, 6, or 8. It is not divisible by 2.

B) Yes. 19,005 is divisible by 3 since the sum of the digits (1 + 9 + 0 + 0 + 5 = 15) is divisible by 3.

C) Yes. Numbers ending in 0 or in 5 are divisible by 5.

236. One gallon of cream. Fold in 5 ounces of whipped cream.
7.9 gallons of cream would require how many ounces of whipped cream?

The conversion factor will be either $\dfrac{1 \text{ gallon}}{5 \text{ oz.}}$ or $\dfrac{5 \text{ oz.}}{1 \text{ gallon}}$

$\dfrac{7.9 \text{ gallons}}{1} \times \dfrac{5 \text{ oz.}}{1 \text{ gallon}} = 39.5$ ounces of whipped cream

237.

	$2x + 5 + 4x$	$= 15x$
Combine like terms	$6x + 5$	$= 15x$
Subtract 6x from both sides	5	$= 9x$
Divide both sides by 9	$\dfrac{5}{9}$	$= x$

| 240–242 | Complete Solutions and Answers |

240. Junior let Janice ride his bike. It was a mile (5,280 feet) to the hospital. He would jog. She got there in 440 seconds. That was 220 seconds before he did. How fast was each of them traveling?

That's a lot of information. The hardest part is to figure out what to do first.

One approach might be to notice that if it took her 440 seconds and she got there 220 seconds before he did, then it must have taken him 660 seconds.

Janice took 440 seconds to go 5,280 feet. Suppose it took her 2 seconds to go 8 feet.* Then she would have been going at 4 feet per second. We divided.

So Janice's speed on the bicycle was $5280 \div 440 = 12$ feet per second.

Similarly, Junior's jogging speed was $5280 \div 660 = 8$ feet per second.

241. On your second try you knock down the other nine pins. What percent of the pins remain standing?

You have knocked down all of the pins.
Zero is what percent of 10?
$0 = ?\%$ of 10

$0 \div 10 = \dfrac{0}{10} = 0 = 0\%$

zero pins remain standing

242. $A \cup B = ?$

A is the set of things Senior watches, which is everything, and B is the set of things Junior watches, which are movies made in the 1940s.

$A \cup B$ means the set of things that are in either A or B (or both).

$A \cup B$ is the set of everything on television. $A \cup B = A$.

---

* Using the **General Rule**, which suggests that you use really simple numbers and see what operation you are using.

Complete Solutions and Answers | 250–252

250. Five ounces is what percent of 3 pounds? Give your answer to the nearest percent.

We need to work either in pounds or in ounces. If we work in pounds, then 5 ounces will be $\frac{5}{16}$ pounds. It will be nicer if we work in ounces.

$$\frac{3 \text{ pounds}}{1} \times \frac{16 \text{ ounces}}{1 \text{ pound}} = 48 \text{ ounces.}$$

5 ounces is what percent of 48 ounces?

$5 = ?\%$ of 48

$\frac{5}{48} \approx 0.1042 = 10.42\% \doteq 10\%$

The box of Ralph's Litter had lost about 10% of its weight.

251. You were two-fifths of the way through your lunch and had eaten 347 Calories. At that rate, how many Calories would you be having for lunch?

347 is $\frac{2}{5}$ of your whole lunch.

$347 = \frac{2}{5}$ of ?

You don't know both sides of the *of* so you divide the number closest to the *of* into the other number.

$$347 \div \frac{2}{5} \quad \Rightarrow \quad 347 \times \frac{5}{2} \quad \Rightarrow \quad 867.5 \text{ Calories}$$

If you were using a calculator,
you punch in ③ ④ ⑦ and then hit the ⊠ key,
then punch in ⑤, then punch the ⑦ key,
then punch in ②, then punch the ⊟ key.

If nothing happens, you probably forgot to turn on your calculator.

252. Jan said, "Tickets are normally $20, but as a cast member I can get a ticket for you at 20% off." How much will you be paying?

If they are 20% off, then you will be paying 80% of the regular price.

80% of $20 = 0.8 \times 20 = 16$. You will be paying $16.

| 253–255 | **Complete Solutions and Answers**

253. Cassie had short eyelashes. Must one of her great-great-grandparents have had short eyelashes?

    No. They all could have been SL and had long eyelashes.

254. Find the LCD of $\frac{1}{2}$ and $\frac{1}{3}$ and $\frac{1}{4}$ and $\frac{1}{5}$

    For the 2 and the 3, it will be **6**.

    For 6 and 4, it will be **12**.

    For 12 and 5, it will be **60**.   *That's one way to do the problem.*

255.   Let set A = the set of Jan's cats that weigh more than 5 pounds.
        Let set B = the set of Jan's cats that weigh less than 8 pounds.
        Let set C = the set of Jan's cats that weight more than 6 pounds.

Question 1: Does A ∪ B equal the set of all of Jan's cats?

    Yes. Every cat that Jan owns must either weigh more than 5 pounds or weigh less than 8 pounds (or both).

Question 2: Is A a subset of B?

    That is the same as asking, "Does every cat that weighs more than 5 pounds be a cat that weighs less than 8 pounds?"

    No. A 10-pound cat would be a member of set A, but not a member of set B. So set A could not be a subset of set B.

Question 3: Is C a subset of A?

    Yes. If it weighs more than 6 pounds, then it must weigh more than 5 pounds.

Question 4: If set D is a subset of set E and set E is a subset of set F, must it always be true that D is a subset of F?

    Yes. If Charlie is in set D and D is a subset of E, then Charlie must be in E. And if Charlie is in E and E is a subset of F, then Charlie must also be in F.

    So every element of D must be in F. That's the definition of subset.

Complete Solutions and Answers | 256–258

256. How many terms in each of these expressions?

$98967x - x$ __2__ The proof: $98967x$ ▌ x

$5 + 4(x^2 - y^2)$ __2__ The proof: 5 ▌ $4(x^2 - y^2)$

$(6 + 6 + 6)^8$ __1__ If I try to put a big black bar in place of a + sign, I would get $(6$ ▌ $6 + 6)^8$ but $(6$ doesn't make sense.

257. What number(s) make $0x = x$ true?

Let's try some numbers.
  If $x = 5$, then $0(5) = 5$ is false.
  If $x = 1$, then $0(1) = 1$ is false.
  If $x = 39769293$, then $0(39769293) = 39769293$ is false.
  If $x = 0$, then $0(0) = 0$ is true.
The only number that makes $0x = x$ true is $x = 0$.

258. How does your body send 15 blood pressure to your lungs and 120 blood pressure to the rest of your body?

You don't have *a* circulatory system. You have two of them.
Hold it! I, your reader, object. You told me that the aorta is the biggest artery in the body. It comes directly out of the heart. How can you have two circulatory systems? Isn't the blood pressure out of the aorta at 120?
Yes. I didn't mention that there are two arteries that come out of the heart. You know about the aorta artery. The other artery is the pulmonary (lung) artery.
Half of your heart pumps 120 blood pressure into the aorta. The other half of your heart pumps 15 blood pressure toward your lungs.

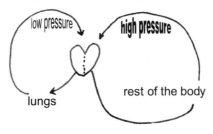

| 260–263 | Complete Solutions and Answers

260. Nine hundred is what percent greater than 35?

    There are 865 more DS babies.   (900 – 35)
    865 is what percent of 35?
    865 = ?% of 35
    865 ÷ 35 ≈ 24.714 = 2471.4% ≐ 2471%

That's a large increase. On the other hand, as you worked out in the *Your Turn to Play* problem #1, a forty-five-year-old mom has a 97% chance (100% – 3%) chance of <u>not</u> having a DS baby.

261. The girl resold her 17¢ polar bear sticker to her friend for $3. What percent gain did she have? (Round your answer to the nearest percent.)

    Working in cents she gained 283¢.
    283 is what percent of 17?
    283 = ?% of 17
    $283 \div 17 = \frac{283}{17} \approx 16.647 \doteq 1665\%$

1665% is not a bad percentage gain.

262. Your blood starts at the heart, goes through the arteries and then into the tiny capillaries and delivers two good things to the cells. Name them.

    The two good things are food and oxygen. (This was mentioned on page 117 of *Life of Fred: Pre-Algebra 1 with Biology*.)

263. How far did the movie producer run?

    So far we have:
        Let t = the number of seconds that they ran.
        Then 12t = the number of feet that Jan ran.
Since we know that the movie producer was running at 9 feet per second, our third line will be:
        Then 9t = the number of feet that the movie producer ran.

Complete Solutions and Answers | 272–274

272. Suppose set E = everything that you have ever seen or touched.
Let set F = everything that your best friend, Jan, has every seen or touched.
   Does the union of E and F equal the entire physical universe?

   No. Neither you nor your best friend Jan have ever seen my sticker collection.

   Are E and F disjoint?
   No. If Jan is your best friend, you both must have seen many things together.

This is a Venn diagram of two sets that are not disjoint:

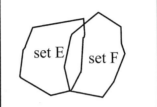

273. Three of Charlie's bites plus 0.7 pounds of his spit weigh a total of 6.1 pounds. How much is each of Charlie's bites?

   You start solving a word problem by letting x equal the thing you are trying to find out.
   Let x = the weight of one of Charlie's bites.
   Then 3x = the weight of three of Charlie's bites.
   Then 3x + 0.7 = the total weight.
At this point the equation becomes easy to write. $3x + 0.7 = 6.1$
   Subtract 0.7 from both sides $3x = 5.4$
   Divide both sides by 3 $x = 1.8$
One of Charlie's bites will take 1.8 pounds out of your ankle.

274. Are you an autotroph?

   That means, "Do you take H_2O, CO_2, and sunlight and make sugars, starches, and oil?
   That means do you do photosynthesis?
   That means, "Are you a plant?"
   By this time, you have probably figured out that you are not an autotroph.

| 275–277 | Complete Solutions and Answers

275. (trickier question) Is blinking your eyes voluntary or involuntary?

Most of the time you walk around and your eyes blink involuntarily. You don't have to think about it. Your eyes stay nice and juicy without any conscious thought on your part.

However, you can seize* control over your blinking. You can tell your eyes to blink.

Blinking is mostly involuntary, but you can make it voluntary.

276. Jan is paid $200 per hour for her singing and dancing. The $73 that she just spent on shirts and a ribbon is what percent of an hour's work?

73 is what percent of 200?
73 = ?% of 200
We don't know both sides of the *of* so we divide the number closest to the *of* into the other number.

$$200 \overline{)73.000} \quad 0.365$$

0.365 = 36.5%
She spent 36.5% of an hour's work to buy that stuff.

277. The density of aNchoVY-LaMB With a RiBBoN oF YaM ice cream is 0.09 pounds per cubic inch. How many pounds does a carton of that ice cream weigh? Round your answer to the nearest pound.

A density of 0.09 pounds per cubic inch means that the conversion factor will be either $\dfrac{0.09 \text{ pounds}}{1 \text{ cubic inch}}$ or it will be $\dfrac{1 \text{ cubic inch}}{0.09 \text{ pounds}}$

From the previous problem we know that the volume is 31.68 cubic inches. We want to change that volume into pounds.

$$\dfrac{31.68 \text{ in}^3}{1} \times \dfrac{0.09 \text{ pounds}}{1 \text{ in}^3} \quad \text{We chose the conversion factor so that the units would cancel.}$$

$$= \dfrac{31.68 \cancel{\text{in}^3}}{1} \times \dfrac{0.09 \text{ pounds}}{1 \cancel{\text{in}^3}}$$

= 2.8512 pounds
≐ 3 pounds   ≐ means equals after rounding.

---

* *Seize* doesn't even follow the complete spelling rule: I before E, except after C, except when pronounced as A in neighbor or sleigh.

Complete Solutions and Answers 278–279

278. Can a solipsist feel pain?

   Solipsists believe that they are the only real thing in the universe. They are the supremely self-centered individual. You break your bones and that's merely entertainment for them.

   But solipsists—who go around thinking that the real world ends at the border they call their skins—are in real pain when they have a toothache or the sniffles or break a bone. Their whole universe is falling apart.

   By contrast, consider the mom and her six kids living out on a farm in Kansas. The kids are all playing outside. She sees a tornado coming their way. She runs outside and screams at her kids to get down into the tornado cellar where they will be safe. The older ones obey. The younger ones want to continue playing on the swings. She picks the little ones up and runs with them. She falls and breaks her arm. She picks them up again and gets them to the cellar.

   Once inside she counts them: 1–2–3–4–5–6. She is breathing hard but feeling very relieved. They are all safe.

   This woman is the opposite of a solipsist. Her reality is outside of herself. Her kids matter. One of her older sons says, "Mom. Your arm looks a little funny. Do you have two elbows?"

   The more you live outside yourself, the less you notice the bad stuff that happens to you.

   One-year-olds are perhaps the world's greatest solipsists. When they get a cold, it's the end of the world for them. It's "unendurable pain" and "it's going to last forever."

279. Three seconds for her to stand up. Walking at 4 feet per second. Jan knew it would take a total of 51 seconds before they were at Stanthony's. How far would they walk?

   If it would take a total of 51 seconds and 3 of those seconds were spent in her standing up, then 48 seconds would be spent walking.

   Four feet per second and 48 seconds. Do we add, subtract, multiply, or divide? Using the General Rule, suppose we were going 4 feet per second for 2 seconds. We would be going 8 feet. We multiplied.

   4 feet per second times 48 seconds = 192 feet to Stanthony's.

| 300–302 | **Complete Solutions and Answers** |

300. Suppose that there is a gene for dogs that gives them a curly tail. If T is the dominant gene that creates a curly tail and if t is the recessive gene, then since January has a curly tail, what must his genotype be?

There are three possible cases: TT, Tt, and tt. We can eliminate tt because January has a curly tail. So January must be either TT or Tt.

301. How much farther did Jan run than the movie producer?

The movie producer was 60 feet ahead of Jan. For Jan to catch up with him, she would have to run 60 feet more than he did.

So far we have:
  Let t = the number of seconds that they ran.
  Then 12t = the number of feet that Jan ran.
  Then 9t = the number of feet that the movie producer ran.
And we know that she ran 60 feet more than he did.
Translation: We know that 12t is 60 feet more than 9t.
Translation: We know that 12t = 60 + 9t.

Solving the equation is the easy part. $12t = 60 + 9t$
  Subtract 9t from both sides $3t = 60$
  Divide both sides by 3 $t = 20$
It took Jan 20 seconds to catch the movie producer.

302. Jan's 63 boxes of macaroni and cheese plus a $1.13 plastic bowl cost a total of $32. How much does a box of macaroni and cheese cost?

  You always start by letting x equal the thing you are trying to find.
  Let x = the cost of a box of m & c.
  Then 63x = the cost of 63 boxes of m & c.
  The 63x + 1.13 = the total cost.
We know the total cost is $32 so the equation is $63x + 1.13 = 32$
The rest is mechanical.
  Subtract 1.13 from both sides $63x = 30.87$
  Divide by sides by 63 $x = 0.49$
A box of m & c costs $0.49.
Or you could write that a box costs 49¢.
What you cannot write is that a box costs 0.49¢.
  0.49¢ is less than a half cent since 0.50¢ is exactly a half cent.

Complete Solutions and Answers | 320–322

320. She drank a two-dollar bottle of cola and washed down 7 purple "throat pills." That's what she called them. The pills and the cola cost $49.46. How much did each pill cost?

You always begin a word problem by letting x equal the thing you are trying to find out.

Let x = the cost of a pill.
Then 7x = the cost of 7 pills.
Then 7x + 2 = the total cost.

At this point you can write the equation: $\quad\quad\quad 7x + 2 = 49.46$

Please do not get in the habit of jumping from the English to the equation. You may be able to do it now, but once you get to the more complicated word problems in algebra, you will be lost without this Let x = *framework. This is just practice for the "big time" to come.*

Subtract 2 from both sides $\quad\quad\quad\quad\quad\quad\quad\quad 7x = 47.46$
Divide both sides by 7 $\quad\quad\quad\quad\quad\quad\quad\quad\quad\quad x = 6.78$

Each purple "throat pill" cost $6.78.

321. $\frac{5}{6} \div \frac{3}{4}$

$= \frac{5}{6} \times \frac{4}{3} = \frac{5}{\underset{3}{\cancel{6}}} \times \frac{\overset{2}{\cancel{4}}}{3} = \frac{10}{9} = 1\frac{1}{9}$

322. She ordered four pieces of chocolate cake with raspberries. The four pieces cost $30.25. The tax was 9%. What was the total bill?

The long way: 9% of 30.25 = 0.09 × 30.25 ≐ $2.72
$\quad\quad$ The cost plus the tax = 30.25 + 2.72 = $32.97

The easier way: 9% plus the original 100% is 109%.
$\quad\quad$ 109% of 30.25 = 1.09 × 30.25 ≐ $32.97

❶ Jan now had $2.23 left. ($35.20 − 32.97) She would be back to macaroni and cheese for lunch and dinner.

❷ When you are a kid, the mistakes you can make are usually less serious than when you are an adult. As a kid you might hold your folk incorrectly. As an adult you might blow all your money on four pieces of chocolate cake with raspberries for breakfast. It is Nature that disciplines adults, and Nature can spank hard.

323–326 Complete Solutions and Answers

323. The way to prove that C is not a subset of J *is to find a member of C that is not a member of J.* That's the easy way to show C is not a subset of J. Must the empty set, { }, be a subset of J?

    Yes. Here's the argument: If { } is *not* a subset of J, then we should be able to find a member of { } that isn't a member of J.
    I look and look and can't find any member of { }.

    The same argument shows that the empty set is a subset of every possible set.

324. Jan had one cat (Juno). She now has two cats. What percent increase did she experience?

    She added one new cat.
    One (the increase) is what percent of one (the original cat)?
    1 = ?% of 1
    We divide the number closest to the other into the other number. You do not (I hope!) need a calculator.

$$1 \overline{\smash{)}\, 1} \;\; \frac{1}{}$$

    1 = 1.00 = 100%
    Jan experienced a hundred percent increase in her cat population.

325. Jan had sold 100 of her $16.35 four hoofprints-plus-album packages last night. How much had she received?

    To multiply by 100 you move the decimal point two places.
    100 × 16.35 = $1635.

326. How many minutes of singing and dancing would it take to buy that stuff?

36.5% of an hour is 36.5% of 60 minutes = 0.365 × 60 = 21.9 minutes
    21.9 minutes looks a bit awkward. It is a correct answer, but I want to play with it a little.
    The 0.9 minutes is 0.9 × 60 seconds = 54 seconds
Jan took 21 minutes 54 seconds to earn those shirts and the ribbon.

Complete Solutions and Answers | 327–331

327. You accidentally stepped on Juno, who had been sleeping after her snack on your ankle. Which of your receptors could tell you what you had done?

    Juno's scream would have activated your audioreceptors. You would have felt pressure (mechanoreceptors) as she sunk her teeth into your other ankle.

328. Cassie had short eyelashes. Must one of her great-great grandparents have had a S gene?

    If none of them had an S gene, then Carrie couldn't have inherited an S gene. So at least one of them had an S gene.

329. Which one is the longest: mouth, pharynx, esophagus, tummy, small intestine, or large intestine?

    I would sure hate to have my mouth as long as my small intestine, which is six or seven yards long. If it were, then my dentist would have to crawl inside my mouth to work on my teeth.

330. The blood went from your heart to your leg muscles through your arteries. It is transported back to your heart through your veins.*

331. For each performance Gordon gets paid $61.32 and Jan gets paid $44. Gordon gets paid what percent more than Jan? Round your answer to the nearest percent.

    Gordon gets paid $17.32 more than Jan.
    $17.32 is what percent of $44?
    17.32 = ?% of 44

You don't know both sides of the *of*, so you divide the number closest to the *of* into the other number.

    $17.32 \div 44 \doteq 0.3936$

    $0.3936 = 39.36\% \doteq 39\%$

Gordon is paid 39% more than Jan.

---

* Do you remember the English rule: *I before E, except after C?*
The word *vein* doesn't seem to follow that rule.

| 333–337 | **Complete Solutions and Answers** |

333. Balance $\qquad\qquad NH_4NO_3 \rightarrow N_2O + H_2O$

There are 4 H's on the left and 2 H's on the right.
Add some H to the right side. $\qquad NH_4NO_3 \rightarrow N_2O + 2H_2O$

The N's are in balance (2 on each side).
The H's are in balance (4 on each side).
The O's are in balance (3 on each side.) ☺

336. Besides getting money to start a business, what are the other parts of a business plan?

  1. Deciding on what product or service you will offer.
  2. Finding out whether people *want* that product or service. It might be silly to try to sell typewriters in today's world.
  3. Finding out how much they might be willing to pay.
  4. Finding out your cost to make that product or provide that service. That would include the cost of the materials, the cost of advertising, the taxes you would face, the regulations you would have to obey, the cost of labor, the cost of office space, and insurance.
  5. Learning what the competition is in your field. If the field is crowded and there is fierce competition, your chances of making a profit are diminished. If a million people are already doing it, consider a different field.
  6. Consider whether you really like making that product or producing that service. You are going to be spending a lot of time—most of your waking hours—doing it. Don't be a doctor if you hate being around sick people. Don't be a mortician if you think that dead people are creepy. Don't be a boxer unless you love getting in the ring (and don't mind getting your face rearranged).

  It should be noted that Janice isn't in love with jams. She just chose it because 𝒥𝒶𝓃'𝓈 𝒥𝒶𝓂𝓈 sounded nice.

  7. Determine how much education you will need in order to be really good in your business. It would be really silly to think of spending years and years learning to be a medical doctor if you are in your seventies.

  If you are not <u>really good</u> in your business, then forget it. Get a job instead. Businesses can make a lot of money—and they can go bankrupt.

337. Describe the 23rd chromosomes of Janice, Junior, Senior, and Janus.

      Janice is female. Hers are XX.
      Junior and Senior are males. Theirs are XY.
      Janus is female. Hers are XX.

Complete Solutions and Answers | 338–341

338. What do you think of this idea? Let's transfer eyeballs to the palms of our hands.

Our eyeballs are currently located in our bony skulls. They aren't hanging out.

They are set deep in our eye sockets with bone all around to protect them from injury.

If our eyes were in our hands, they would be very vulnerable. You are at a performance and everybody gives a great big round of applause—and everyone becomes blind.

339. What is the weight (in grams) of an Avogadro's number of atoms of neon? This is the same as asking how much 6.0221236×10^{23} atoms of neon weigh.

This is the same as asking how much does a mole of neon atoms weigh. This was the previous question. 20.18 grams.

341. Put an "H" (for healthy) or a "U" (for unhealthy) in front of each of these:

    __U__ smoking Once you start smoking it's hard to stop.
    __U__ driving a motorcycle at night in the middle of a snow storm
    __H__ being a mathematician
This is one of the safest occupations.

    __U__ eating a lot of candy
Sugar does a lot of bad things besides rotting your teeth. When you get older and start talking to your internal organs, you should ask your pancreas what it thinks of sugar.

    __H__ exercising
Sitting all day will shorten your life. Running 30 miles a day through the jungle will shorten your life. Between these two extremes is a happy place—not too little and not too much.

    __H__ wearing a seatbelt
When I was a kid, no one wore seatbelts. That's because cars didn't have seatbelts. A girlfriend I knew years ago had a big scar on her nose. When she was young and there weren't seatbelts, she was in a car accident. Her head went through the windshield.

| 345–348 | Complete Solutions and Answers

345. Joe was running at 14 feet per second.
Let t = the number of seconds before the pelican catches Joe.
How far did Joe run before the pelican caught him?

d = rt   becomes   d = 14t   Joe ran 14t feet before he was caught.

346. What about reducing $\frac{12}{732}$ by doing $\frac{1\cancel{2}}{732}$ and getting $\frac{1}{73}$?

The only way we reduce a fraction is to divide top and bottom by the same number.

Taking $\frac{12}{732}$ and striking off the 2 ⇒ $\frac{1\cancel{2}}{732}$ isn't dividing top and bottom by the same number.

$\frac{670}{8330} = \frac{67\cancel{0}}{833\cancel{0}} = \frac{67}{833}$ works. It's dividing top and bottom by 10.

Slashing off zeros works, but that doesn't mean that you can slash off 2's.

$\frac{1\cancel{2}}{732} \neq \frac{12}{732}$

It's easy to prove that $\frac{12}{732}$ doesn't equal $\frac{1}{73}$

Grab a calculator. $\frac{12}{732} \approx 0.0163934$

$\frac{1}{73} \approx 0.0136986$

347. Combine like terms in the equation $4y^2 - y^2 + 17 = 3x^2y + 18xy$

$3y^2 + 17 = 3x^2y + 18xy$

$4y^2 - y^2$ is the same as four y^2 take away one y^2.

You can't combine $3x^2y + 18xy$. Two like terms must have the same variables with the same corresponding exponents.

348. Jan looked at her 25 sparklettes. Sixty percent of them were red. Thirty-two percent of them were green. The rest were yellow. How many were yellow?

8% of them were yellow. (100% – 60% – 32%)
8% of 25 = ?
0.08 × 25 = 2   Two of them were yellow.

Complete Solutions and Answers — 349

349. Name a number that is smaller than −73.

There are many possible answers. I like −75.
−75 < −73 If you added 2 to −75 you would get −73.
Other possible answers are −74, −76, −100, and −1183239389739.
One number that is a little less than −73 is −73.00002.

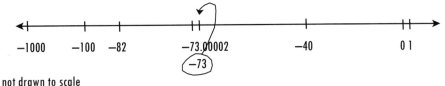

not drawn to scale

For English majors we should note that when you write a negative 75, you attach the minus sign right next to the 75.   −75

When you *do* subtraction, you leave a space on either side of the subtraction sign.   3 − 52 = −49

When you use a dash in a sentence—as I am doing now—you leave no space.

For advanced English majors: Subtraction signs are en dashes. Dashes in sentences are em dashes. en dashes are shorter than em dashes.   en dash –
           em dash —
And hyphens are the shortest of all.   hyphen -

For super advanced English majors: en dashes are also used between years. 1939–1945 We are not subtracting so we don't include spaces.

For extremely super advanced English majors: We use a hyphen to separate words. Denver-Omaha train line. We use an en dash if we are separating pairs of words. San Francisco–New York train line.

There is such a thing as a 2-em dash (also known as a double dash), but most English majors have never heard of it.

155

| 350–360 | **Complete Solutions and Answers** |

350. Solve $\quad\dfrac{w}{3} = \dfrac{7}{8}$

Cross multiply $\quad 8w = 21$

Divide both sides by 8 $\quad w = \dfrac{21}{8}$

Do the arithmetic $\quad w = 2\dfrac{5}{8}$

355. Match each medical name in the first column with its ordinary name.

sternum → breastbone
patella → kneecap
pinna → ear (We did this in Chapter 19.)
cranium → skull
femur → thighbone

I like femurs. They keep my patellas away from my hips. You have two femurs. They are the longest bones in the body and also the strongest.

359. He had run 1.2 miles to the hospital at 8 mph (miles per hour). How many hours did it take him? How many minutes was that?

Let t = the number of hours it took Junior to get to the hospital.
d = rt becomes $\quad 1.2 = 8t$
Divide both sides by 8 $\quad 0.15 = t$

It took Junior 0.15 hours. To convert that into minutes we use a conversion factor.

$$\dfrac{0.15 \text{ hours}}{1} \times \dfrac{60 \text{ minutes}}{1 \text{ hour}} = 9 \text{ minutes}$$

360. The location of a gene on a chromosome is called the <u>locus</u> of that gene.

Complete Solutions and Answers 366–368

366. Junior has long eyelashes. His sister, Ivy, has short eyelashes. His mom, Janus, has short eyelashes. Two questions: Does Jan Senior have long eyelashes? What is the genotype of Junior?

We know that Ivy has SS as her genotype because L is dominant. If she had L, she would have long eyelashes.

Janus also has SS for the same reason.

Since Junior has long eyelashes, he must have at least one L. He couldn't have gotten it from Janus, so he must have gotten it from his father. So Jan Senior has at least one L. Therefore Jan Senior must have long eyelashes.

Junior got an L from his father and an S from his mother.

367. Which of these has Jan failed to consider?
A) Ivy knows a lot about the ice cream business. She has been doing this for years. Jan knows nothing about jams.
B) It takes some saved-up money to open a new business.
C) Is there a market for jams? We know that lots of people like two-dollar pies and 42¢ scoops of ice cream.

Jan has failed to consider any of these things. Her chances of success in creating **Jan's Jams** is low.

368. At Stanthony's it was Hot Chocolate Day: 25% off all hot chocolate drinks. After the discount it would only cost $1.95. What was the price before the discount?

Let's let x = the regular price.

Since there is a 25% discount, Jan would be paying 75% of the regular price.

$1.95 is 75% of x

You don't know both sides of the *of*, so you divide the number closest to the *of* into the other number.

$1.95 \div 75\% = 1.95 \div 0.75 = \$2.60$

We could check our answer by starting with the $2.60 and taking a 25% discount, which is 65¢ (= 2.60 × 0.25). Then the discounted price would be $1.95 (= 2.60 − 0.65). It checks.

369–371 Complete Solutions and Answers

369. Jan had worked for four hours per night for 21 nights and had made $420. Using conversion factors determine how much this actress made per hour.

Start with $\dfrac{\$420}{\text{the whole job}}$

$$\dfrac{\$420}{\text{the whole job}} \times \dfrac{\text{the whole job}}{21 \text{ nights}} \times \dfrac{1 \text{ night}}{4 \text{ hours}}$$

$$= \dfrac{\$420}{\cancel{\text{the whole job}}} \times \dfrac{\cancel{\text{the whole job}}}{21 \cancel{\text{ nights}}} \times \dfrac{1 \cancel{\text{ night}}}{4 \text{ hours}}$$

$= \$5 \text{ per hour}$

370. Jan has 8 pounds of rings on her hand. If she is wearing 13 rings, how much does each ring weigh? (Assume all the rings weigh the same. Round your answer to the nearest ounce. 1 pound = 16 ounces.)

We know 1 pound = 16 ounces.
That gives us the conversion factor.

Start with 8 pounds. $\dfrac{8 \text{ pounds}}{1}$

We need to change pounds into ounces.

$$\dfrac{8 \text{ pounds}}{1} \times \dfrac{16 \text{ ounces}}{1 \text{ pound}} = 128 \text{ ounces}$$

If the 13 rings weigh 128 ounces, how much does each ring weigh? Using the **General Rule**, suppose we had 3 rings that weighed a total of 12 ounces. Then each ring would weigh 4 ounces. We divided.

So 128 ounces ÷ 13 ≈ 9.846 ≐ 10 ounces
Those are heavy rings!

371. If someone turned in ten gray cats, how much reward might they get?

If Jan had wanted to offer a reward of fifty cents, she should have written 50¢, not .50¢
.50 is equal to one half. .50¢ is a half cent. Ten half cents is equal to 5¢, which is a nickel.

Lost cat!!!!
It's gray.
Reward!!!!!
.50¢

Complete Solutions and Answers 372–374

372. Is breathing voluntary or involuntary?

    It's the same as blinking—usually it is involuntary, but you can make it voluntary.

    This is double nice.

① It is nice that it is involuntary. You can go to sleep at night and not worry about remembering to breathe.

② It is nice that you can control your breathing when you want to. Imagine how bad it would be if you were playing in a swimming pool and you decided to swim underwater and you couldn't tell your lungs to stop breathing.

    Imagine how hard it would be to blow up a balloon if you couldn't control your breathing.

    What would it be like to sing if you couldn't control your breathing. There is a special word for when this happens.*

373. What two numbers make $5x = x^2$ true?

    Let's try some numbers.

        If $x = 1$, then $5(1) = 1^2$ is false.
        If $x = 2$, then $5(2) = 2^2$ is false.
        If $x = 5$, then $5(5) = 5^2$ is true.
        If $x = 7$, then $5(7) = 7^2$ is false.
        If $x = 0$, the $5(0) = 0^2$ is true.

If x equals either 5 or 0, then $5x = x^2$ is true.

374. The pelican was running at 16 feet per second.
Let t = the number of seconds before the pelican catches Joe.
How far did the pelican run before it caught Joe?

    $d = rt$ becomes $d = 16t$

The pelican ran for 16t feet before it caught Joe.

    (Or you could say that the pelican ran $14t + 100$ feet, since he ran 100 feet farther than Joe.)

---

* It is called snoring = "singing" in your sleep.

377–378 Complete Solutions and Answers

377. Joe was full. Really full. He told Darlene, "I won't be able to eat anything until lunch." It was 10:47. How long will it be until noon?

From 10:47 to 11:00 is   13 minutes
From 11:00 to noon is   1 hour

It would be one hour and thirteen minutes until Joe would be ready to eat again.

378. Balance $\quad CH_3NH_2 + O_2 \rightarrow CO_2 + H_2O + N_2$

The first giant hint was "Start by balancing the letter that appears in the fewest molecules in the equation."
Let's add a N to the left side. $2CH_3NH_2 + O_2 \rightarrow CO_2 + H_2O + N_2$

There are 10 H's on the left side and 2 H's on the right side. $2CH_3NH_2 + O_2 \rightarrow CO_2 + 5H_2O + N_2$

There are 2 O's on the left and 7 O's on the right. This is tough. Make both sides have 8 O's. $2CH_3NH_2 + 4O_2 \rightarrow CO_2 + 6H_2O + N_2$

2 C's on the left; 1 C on the right. $2CH_3NH_2 + 4O_2 \rightarrow 2CO_2 + 6H_2O + N_2$

The N's balance (2 on each side). The C's balance (2 on each side). The O's don't balance. The H's don't balance.

Hold it! I, your reader, think that this is crazy. You balanced the N's. You balanced the H's. Then the O's. Then the C's. Now the O's and the H's are out of whack again. Will this go on forever?

I hope not.

What do you mean, "I HOPE not"? Or should I have written, "I HOPE not."? with a period and a question mark?

I'm not sure about the period and the question mark business. My guess is the way you first wrote it (without the period) is better English.

My guess is that balancing these chemical equations won't fall into an infinite loop.

(continued on next page)

Complete Solutions and Answers | 380

small essay
Solving Algebra Equations vs. Stoichiometry

With stoichiometry (balancing chem equations), we start out with an unbalanced equation. We adjust the various coefficients until everything becomes balanced.

With algebra equations, we start out with a balanced equation, such as $6x + 7 = 25$. Then we do the same thing to both sides, which keeps things balanced. $6x + 7 = 25$ becomes $6x = 18$, which becomes $x = 3$.

With algebra we have a definite approach. With stoichiometry we have to sometimes just thrash around till it gets balanced.

end of small essay

We now have 10 H's on the left and 12 H's on the right.
$$2CH_3NH_2 + 4O_2 \rightarrow 2CO_2 + 6H_2O + N_2$$

My thinking at this point: If I change the $2CH_3NH_2$ to $3CH_3NH_2$ I will have 15 H's on the left side and I could never adjust the $6H_2O$ to equal an odd number.

I make both sides with 20 H's.
$$4CH_3NH_2 + 4O_2 \rightarrow 2CO_2 + 10H_2O + N_2$$

Make both sides with 4 C's.
$$4CH_3NH_2 + 4O_2 \rightarrow 4CO_2 + 10H_2O + N_2$$

Add a couple more N's to the right side.
$$4CH_3NH_2 + 4O_2 \rightarrow 2CO_2 + 10H_2O + 2N_2$$

8 O's on left; 14 O's on the right.
$$4CH_3NH_2 + 9O_2 \rightarrow 2CO_2 + 10H_2O + 2N_2$$

I can't believe it: C's match (4 on each side). H's match (20 on each side). N's match (4 on each side). O's match (18 on each side). This deserves a big happy face. ☺

380: How much farther did the pelican run than Joe before it caught Joe?

Since Joe had a 100-foot head start, the pelican had to run 100 feet more than Joe in order to catch up to Joe.

Complete Solutions and Answers

409. Ivy sold $\frac{4}{5}$ of a gallon of Buffalo Puffing this last month. She has sold $\frac{7}{10}$ of a gallon Sneaky Dog this last month. Which flavor has sold less?

Which is smaller? $\frac{4}{5}$ or $\frac{7}{10}$

To compare two fractions, you pretend that you are going to add them. You give them the same denominators. Then it is easy to tell which is smaller.

$$\frac{4}{5} + \frac{7}{10} = \frac{4 \times 2}{5 \times 2} + \frac{7}{10} = \frac{8}{10} + \frac{7}{10} \qquad \frac{7}{10} \text{ is smaller.}$$

Sneaky Dog sold less.

410. Your lunch bill was $4.89. Jan's was $11.22. Jan's bill was what percent greater than yours? Round your answer to the nearest percent.

Jan's bill was $6.33 more than yours. (11.22 − 4.89 = 6.33)
$11.22 is an *increase* of $6.33 over your $4.89.
6.33 is what percent of 4.89?
6.33 is ?% of 4.89

We don't know both sides of the *of* so we divide the number closest to the *of* into the other number.

$$\begin{array}{r} 1.294 \\ 4.89 \overline{)6.330} \end{array}$$

The numbers are much worse than in the previous book, but we can use calculators now. I divided to the third decimal place so that I could round to the nearest percent.

1.294 = 129.4% ≐ 129%   ≐ means "equals after rounding"

Jan's lunch was 129% more expensive than yours.

411. She already has 16 cats. If she gets one more, what percentage increase will that be?

The increase will be one cat.
1 is what percent of 16?
1 = ?% of 16
$1 \div 16 = \frac{1}{16} = 0.0625 = 6.25\%$

Six and one-fourth percent is not a large percentage. She will hardly notice the difference.

Complete Solutions and Answers | 412–414

412. Which is smaller: $\frac{5}{6}$ or $\frac{3}{4}$?

To compare two fractions we pretend that we are going to add them. We make the denominators (the bottoms alike).

$$\frac{5}{6} = \frac{10}{12} \qquad \frac{3}{4} = \frac{9}{12}$$

Then it is easy to see that $\frac{3}{4}$ is smaller. $\qquad \frac{3}{4} < \frac{5}{6}$

413. Cassie had short eyelashes. Must at least two of her great-great grandparents have had an S gene?

If only one of her great-great grandparents had an S gene, then only one of her eight great grandparents could have had an S gene.

Then only one of her four grandparents could have had an S gene.

Then only one of her two parents could have had an S gene.

Then Cassie could have inherited only one S gene. But she has short eyelashes. Her genotype is SS.

Therefore, she must have had at least two great-great grandparents with an S gene.

414. Are the set of words Gordon speaks in the play and the set of words that Jan speaks in the play disjoint?

That is the same as asking if the two sets have nothing in common. Since Jan is saying nothing in the play, she and he don't both utter any particular word. The sets are disjoint.

The set of words that Jan utters is called the empty set. It is sometimes written as { }.

The set of words that Gordon speaks is {Hee, Haw}.

The empty set is disjoint from every other set that you can think of.

Two facts:

① If S is any set, { } and S are disjoint.
② If S is any set, { } ∪ S = S.

| 424 | Complete Solutions and Answers |

424. What's two times zero?

$2 \times 0 = 0$ and 0×2 also equals 0.
Zero times any number equals zero.
In algebra we will write $0x = 0$ for any x.

Without a calculator you can figure out the final answer for:

$$\frac{4.398^2 - 8,000}{3(\pi - 2) + 5} \times \log(\sqrt{6}) \times \sin 44° \times e^7 \times 0$$

Wait a minute! I, your reader, understand that multiplying by zero will give you an answer of zero. That's easy. I've got three questions. What's log? What's sin? What's e?

log is a function that we'll study in advanced algebra. Just briefly, note that $\log 10^3$ is 3 and $\log 10^7$ is 7 and $\log 10^\pi$ is π.

We'll study the sin function in trig. It deals with triangles. Briefly, $\sin 30° = \frac{1}{2}$.

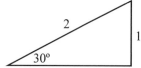

e is just a number like π is a number.
$\pi \approx 3.14159265358979323846264338 32795$
$e \approx 2.71828182845904523536028747 13527$

I know what pi (π) is used for. The diameter of a circle times pi always equals the distance around the circle (the circumference). But I've never seen e before. What's that all about?

Ahead of you is . . .
  Beginning algebra (solve $29x + 3 = 90$).
  Advanced algebra (find $\log 10^9$).
  Trig (find $\sin 52°$).

And then comes college calculus in which we will need e. In calculus we find the areas under curves.
  Here is the curve $y = x^2$. You will learn to draw this in beginning algebra.

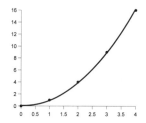

In calculus we will find the area under that curve between x = 2 and x = 3.5. In calculus we make it look complicated by writing that area as $\int_{x=2}^{3.5} x^2 \, dx$.

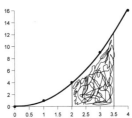

That looks horrible, but it is easy to calculate. (I could show you in less than a minute, if I had the time.)

Now you know enough calculus to write down what the area under the curve $y = 6x^{13} + 5x$ between x = 5 and x = 7 is.

$$\int_{x=5}^{7} (6x^{13} + 5x) \, dx$$

The number e will come in handy when we do $\int_{x=1}^{8} \frac{1}{x} \, dx$.

When you were 12 months old, walking across the room was as hard as your doing a 10-mile walk today.

When you were in first grade, $\frac{3}{4}$ looked mysterious. Back then understanding what $\frac{3}{4}$ meant was as hard as figuring what 38% of 7 is now. And when you get to the second semester of calculus it will be just as hard to do $\int \cos x \, dx = \sin x + C$. (sin, cos, and tan are trig functions.) When you have finished calculus, all this will look dorky easy.

425. Janus was in the hospital with a broken femur. What's a femur?

The femur is the one bone in the upper leg. Some people call it the thighbone.

| 430–432 | **Complete Solutions and Answers** |

430. Ivy bought a giant ice cream cone to place outside her building. The cone is 50 feet tall and 20 feet across at the top. What is its volume? (Round your answer to the nearest cubic foot.)

The volume of a cone is $\frac{1}{3}\pi r^2 h$,
where r is the radius of the cone and
h is the height of the cone.

If the cone is 20 feet across, then the radius is 10 feet.

$V_{cone} = \frac{1}{3}\pi r^2 h$ becomes	$\frac{1}{3}\pi(10^2)(50)$
Doing the arithmetic	$1,666\frac{2}{3}\pi$
Letting $\pi = 3.1416$	$1,666\frac{2}{3} \times 3.14156$
Doing the arithmetic	5235.999
Rounding to the nearest cubic foot	$5,236$

The volume of Ivy's cone is 5,236 cubic feet.

431. $12 for 4 minutes of bowling. How much is that per hour? (Use a conversion factor.)

$$\frac{\$12}{4 \text{ minutes}} \times \frac{60 \text{ minutes}}{1 \text{ hour}}$$

$$= \frac{\$12}{4 \text{ minutes}} \times \frac{60 \text{ minutes}}{1 \text{ hour}}$$

$$= \frac{\$180}{1 \text{ hour}} \quad \text{At \$180 per hour, bowling can be pretty expensive.}$$

432. If set M is a subset of set N, then what must M ∪ N equal?

Let's experiment.
Suppose M = {1, 2, 3} and N = {1, 2, 3, 4, 5}.
Then M is a subset of N.
Then M ∪ N = {1, 2, 3, 4, 5}.
So M ∪ N = N.

If you like pictures

M is a subset of N
The set of everything that is in either M
or in N (M ∪ N) is the set of everything in N.

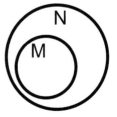

166

Complete Solutions and Answers | 433–436

433. Charlie is an obligate carnivore. Only living things can do photosynthesis. Can Charlie do photosynthesis?

    No. Photosynthesis takes water and carbon dioxide and energy and produces sugar, starch, oil, and oxygen.

    What Charlie does is just the opposite. He takes food and oxygen and produces water, carbon dioxide and energy.

434. Solve $3x + 1.9 = 10$.

We start with	$3x + 1.9 = 10$
Subtract 1.9 from both sides	$3x = 8.1$
Divide both sides by 3	$x = 2.7$

435. x squared is x^2. x cubed is x^3.

    What is forty-four x cubed times y to the seventh power?

    $44x^3y^7$ Note that "forty" doesn't have a u in it.

436. Jan had $193.44 in her pocket. How long would her money last if she spent it at the rate of $5 every 4 minutes?

    We will be converting $193.44 into minutes of shopping.

    Since $5 matches up with 4 minutes, the conversion factor will be either $\dfrac{\$5}{4\text{ min}}$ or $\dfrac{4\text{ min}}{\$5}$

$$\frac{\$193.44}{1} \times \frac{4\text{ min}}{\$5} = \frac{773.76}{5} \text{ minutes} \approx 154.75 \doteq 155 \text{ minutes},$$

which is 2 hours and 35 minutes.

To divide 773.76 by 5, you could either us a calculator or divide it out by hand. Doing it by hand takes about 18 seconds.

```
        154.75
    5)773.76
      5
      —
      27
      25
      —
       23
       20
       —
        37
        35
        —
         26
         25
         —
```

167

| 437–439 | **Complete Solutions and Answers** |

437. A person's femur is about 27% of their height. This is true for both men and women.

Some anthropologist was digging in some field in Kansas and found a thighbone of someone who died a zillion years ago. It was 18 inches long. How tall (to the nearest inch) was that person?

18" is 27% of total height
18 = 27% of height
18 ÷ 0.27 ≈ 66.6666 ≐ 67 inches   (= 5' 7" tall)

438. Joe's most recent blood pressure reading was 139/87. What is his systolic reading?

From Chapter 24, we learned that the upper number in a blood pressure reading is call the systolic reading. Joe's is 139.

439. *Roma hot chocolate* comes in a large cup—1,200 ml. How many quarts is that? (1,000 ml = 1 liter. 1 liter = 1.06 quarts.)

$$\frac{1200 \text{ ml}}{1} \times \frac{1 \text{ liter}}{1000 \text{ ml}} \times \frac{1.06 \text{ quarts}}{1 \text{ liter}} \quad \text{(two conversion factors)}$$

$$= \frac{1200 \text{ ml}}{1} \times \frac{1 \text{ liter}}{1000 \text{ ml}} \times \frac{1.06 \text{ quarts}}{1 \text{ liter}}$$

= 1.272 quarts

Complete Solutions and Answers 440–443

440. . . . like three-year-olds in a calculus class. Is that a simile or a metaphor?

It is a simile. Similes use *like* or *as*.

441. What percent of the smallest hemoglobin (17,000) molecule is that little iron atom (56)?

Round your answer to the nearest hundredth of a percent.

56 is what percent of 17,000?
56 = ?% of 17,000
56 ÷ 17,000 ≈ 0.0032941176470588235294117647 0588
   = 0.32941176470588235294117647 0588%
   ≈ 0.33%

If your calculator only read 0.0032941 instead of 0.0032941176470588235294117647 0588, you would still have arrived at the same answer as I did.

If you had just used paper and pencil

```
          0.00329
17000)56.00000
      51000
       50000
       34000
      160000
      153000
```

you would still have arrived at the same answer I did.

442. Make a guess. When will we divide by zero?

On days that don't end in "y."

443. Aren't all dogs born in January named Jan?

| 446–452 | Complete Solutions and Answers |

446. *Make a guess.* Why wasn't Ivy at her ice cream store?

Let's look at what you know:
① The hospital had phoned Junior to tell him his mom had a stroke.
② Ivy was Junior's sister.
③ Ivy also lived nearby.

It wouldn't be surprising that they had also called her. Even if they hadn't. Junior would have called Ivy. Ivy headed to the hospital to see her mom.

449. If you were to take a heart cell from January and analyze the set of genes in that cell, could you tell whether January was TT or Tt?

Virtually every cell in January's body contains all the genes. It seems a bit strange, but in January's heart cells are all the instructions for making every part of his body.

We know that no two people have the same fingerprints. One's genotype is even more distinctive (except for identical twins). I refuse to say that it's "more unique" because every English major knows that *unique* is an absolute adjective.

452. If she had saved $130 per month and had earned 0.9% on her money each month, at the end of six years (72 months) she would have

$$130 \times \frac{(1+0.009)^{72} - 1}{0.009}$$

I'll do the hard part: $1.009^{72} \doteq 1.906$

Find out how much Jan would have at the end of six years.

$$130 \times \frac{(1+0.009)^{72} - 1}{0.009} = 130 \times \frac{1.906 - 1}{0.009}$$

$$\doteq 130 \times 100.66666$$
$$\doteq \$13{,}087.$$

Without earning interest, she would have only had $9,360 (= 72 × 130)

Complete Solutions and Answers | 454–457

454. Every sticker in my sticker collection was either given to me or I bought it at the store. If set G = those stickers that have been given to me and if set H = those stickers I bought at the store, then is the union of G and H equal to my sticker collection? Are G and H disjoint?

    The union of G and H does equal my sticker collection. (That's not true of my son's sticker collection. He sometimes makes his own stickers.)

    G and H are disjoint. No sticker can both be one that has been given to me and at the same time be one that I bought at the store. The union of sets G and H is sometimes written as G ∪ H.

455. Before getting this gray cat Jan had 35 scratches on her face and arms. She now had 20% more. How many scratches does she now have?

    20% plus the original 100% is 120%.
    120% of 35
    $1.2 \times 35 = 42$ scratches Jan now has

456. People would happily trade 7 cartons of 𝕒ℕ𝒸𝒽o𝕍𝐲-𝕃𝕒𝕄𝔹 𝐰𝐢𝐭𝐡 𝕒 ℝ𝕚𝔹𝔹o𝕟 o𝔽 𝕐𝕒𝕄 ice cream for 3 cartons of strawberry ice cream. If you had 623 cartons of the smelly stuff, how many cartons of strawberry ice cream could you trade it for?

The conversion factor will be either $\dfrac{3 \text{ strawberry}}{7 \text{ anchovy}}$ or $\dfrac{7 \text{ anchovy}}{3 \text{ strawberry}}$

Note that the numerator and denominator are equal. That makes the fraction equal to one. When you multiply by a conversion factor, you are multiplying by one, and that doesn't hurt anything.

    We are starting with 623 cartons of the anchovy.

$$\dfrac{623 \text{ anchovy}}{1} \times \dfrac{3 \text{ strawberry}}{7 \text{ anchovy}} \quad \text{We chose the conversion factor so that the anchovies would cancel.}$$

$$= \dfrac{623 \text{ \cancel{anchovy}}}{1} \times \dfrac{3 \text{ strawberry}}{7 \text{ \cancel{anchovy}}}$$

$$= 267 \text{ cartons of strawberry ice cream}$$

457. $87 – $100 – $120. That's the same as 87 – 220. What does that equal?

    A gain of 87 and a loss of 220 equals –$133. Jan is losing money!

171

| 458–459 | Complete Solutions and Answers |

458. If she started at 118 pounds and dropped to 100 pounds, what percent would she have lost? Give your answer to the nearest percent.

> She would have dropped 18 pounds.
> 18 pounds is what percent of her original 118 pounds?
> 18 is ?% of 118

We do not know both sides of the *of* so we divide the number closest to the *of* into the other number.

> $18 \div 118 = 0.15254237288135593220338983050847$

(My calculator is bigger than yours. ☺)

> $0.15254237288135593220338983050847 \doteq 0.15 = 15\%$

> $\doteq$ means "equals after rounding."

Having a big calculator isn't any advantage in this problem. All we needed to have is $18 \div 118 = 0.152$ so that we could round to the nearest percent. In fact, my having to write out 0.15254237288135593220338983050847 was just showing off. I apologize.

I, your reader, accept your apology. By the way, I want one of those calculators! My calculator only gives me about eight or ten digits. How much do they cost?

> I got mine for free. **That's a good price!** It came built into my computer. I can do $\sqrt{2} = 1.4142135623730950488016887242097$.

459. Each night Jan arrived at the theater at 6 p.m. The first 45 minutes were spent putting on stage makeup and getting into costume. After the performance itself, it took 30 minutes to take off the makeup and get out of the costume. We already know how much Jan was paid for those four hours. If you have ever done acting, you know that there were other hours that Jan wasn't paid for. What were they?

> There were hours Jan spent learning the lines to be spoken. There were the many hours spent in doing auditions, most of which didn't result in employment.
>
> If you have ever done teaching that was paid by the classroom hour, you know that weren't paid for the hours you spent preparing for your lectures or the hours spent in correcting your students' homework.
>
> If you have ever been a real estate agent and spent 40 hours putting together a sale that resulted in a $2,000 commission (= $50/hour), you know that you weren't really being paid $50/hour because of the many deals that fell through where no commission was paid.

Complete Solutions and Answers | 460–463

460. Stanthony's 16-inch pizza was $5.76. (A 16-inch pizza means that its diameter is 16 inches.) What was the area of this pizza? (Use 3 for π in this problem. $\text{Area}_{circle} = \pi r^2$ where r is the radius.)

If the diameter is 16 inches, then the radius is 8 inches.
$\text{Area}_{circle} = \pi r^2$ becomes $\text{Area}_{circle} \approx (3)(8^2) = 192$ square inches.
($\approx$ means "approximately equal to.")

461. Nine liters equals how many quarts? (1 liter = 1.057 quarts).

The conversion factor will be either $\dfrac{1 \text{ liter}}{1.057 \text{ quarts}}$ or $\dfrac{1.057 \text{ quarts}}{1 \text{ liter}}$

$\dfrac{9 \text{ liters}}{1} \times \dfrac{1.057 \text{ quarts}}{1 \text{ liter}} = 9.513 \doteq 9.5$ quarts

That's a lot of liquid.

462. Before Jan started performing, Ivy was filling 60 ice cream orders each day. After a week, Ivy will filling 230% more orders each day. How many orders is that?

Long way: 230% of 60 is 2.3 × 60, which is 138.
Then add 138 to the original 60 to get 198 orders.

Short way: 230% plus the original 100% is 330%. (That step can always be done in your head.) Then 330% of 60 is 3.3 × 60, which is 198.

463. How long is it from 9:20 to 10:07?

It's 40 minutes from 9:20 to 10:00.
It's 7 minutes from 10:00 to 10:07.

```
  40
+  7
  47 minutes
```

I hope you didn't need a calculator for that.

173

464–467 Complete Solutions and Answers

464. Your blood circulates: heart → arteries → capillaries (where the cells are) → veins → heart. At the cells, food is dropped off. Where was the food picked up in the first place?

    The food (the sugars and amino acids) are picked up through the walls of the small intestine.

465. Darlene drove him to her favorite hospital (Beth's Vet). She drove at 12 mph, and they spent 5 minutes in the parking lot trying to find a space. It was now 11:08. How long were they in the car?

    From the previous problem they started at 10:47. They got out of the car at 11:08.   From 10:47 to 11:00 is   13 minutes.
                          From 11:00 to 11:08 is   8 minutes.
They were in the car for 21 minutes.

466. You walked the 2,000 feet to the theater. You walked at the rate of 4.5 feet per second. How many seconds did it take you? (Round your answer to the nearest second.)

    Let t = the number of seconds it took you to get to the theater.
    $d = rt$ becomes $2000 = 4.5t$
    Divide both sides by 4.5 $444.444 \approx t$
    Round to the nearest second $444 \doteq t$

It took about 444 seconds to get to the theater.

467. Your mother's sister has short eyelashes. Your mother's mother has short eyelashes. Could your mother's father have a genotype of SL?

your grandparents

Your mom   your aunt

If your aunt has short eyelashes, then she received an S from each of your grandparents.

That means that your grandfather must have at least one S. He could be either SS or SL.

Complete Solutions and Answers 469–475

469. Balance $\quad H_2 + N_2 \rightarrow NH_3$

There are 2 N's on the left and 1 N on the right.
Add an N to the right side. $\quad H_2 + N_2 \rightarrow 2NH_3$

There are 2 H's on the left and 6 H's on the right.
Add 4 H's to the left side. $\quad 3H_2 + N_2 \rightarrow 2NH_3$

The H's are in balance (6 on each side).
The N's are in balance (2 on each side). ☺

Making ammonia, NH_3, commercially is done by combining hydrogen gas, H_2, with nitrogen gas, N_2. It's called the Haber process.

470. $\frac{1}{3}$ of a scoop of chocolate, and $\frac{1}{5}$ of a scoop of anchovy-lamb with a ribbon of yam and the rest of the scoop will be vanilla. How much vanilla?

$\frac{1}{3} + \frac{1}{5}$ is $\frac{8}{15}$

That leaves $\frac{7}{15}$ for vanilla.

$$\frac{1}{3} + \frac{1}{5} = \frac{5}{15} + \frac{3}{15} = \frac{8}{15}$$

$$1 - \frac{8}{15} = \frac{15}{15} - \frac{8}{15} = \frac{7}{15}$$

475. Janice said, "I've always wanted to have a number of children that was evenly divisible by 3."

Junior, "I want an even number of kids."

What are the first three possible number of kids that they would both be happy with?

The smallest possible number that both 2 and 3 divide evenly into is 6. The next smallest is 12. The third smallest is 18.

What really becomes fun is if they also try to please Janus. She has always wanted the number of children Junior has to be divisible by 5.

Some uptight people might think that 30 kids is excessive.

485–490 | Complete Solutions and Answers

485. How much (in grams) does a dozen neon atoms weigh?

From problem #339, we know that 6.0221236×10^{23} atoms of neon weigh 20.18 grams.

We want to convert a dozen atoms into a weight in grams.
We know that 6.0221236×10^{23} matches up with 20.18 grams. That gives us the conversion factor.

$$\frac{12 \text{ atoms}}{1} \times \frac{20.18 \text{ g}}{6.0221236 \times 10^{23} \text{ atoms}}$$

$$= \frac{12 \text{ atoms}}{1} \times \frac{20.18 \text{ g}}{6.0221236 \times 10^{23} \text{ atoms}}$$

$$= \frac{(12)(20.18) \text{ g}}{6.0221236 \times 10^{23}}$$

$$= \frac{40.211728 \text{ g}}{10^{23}}$$

$$= 0.0000000000000000000040211728 \text{ grams}$$

Twelve atoms of neon aren't very heavy.

489. Name a number smaller than $\dfrac{\pi}{890,385,227.614}$

The hard way to do this problem is to do the division:
$3.141592653589793238462643383279 \div 890,385,227.614$
$\approx 0.000000003528352174045431465361395156018$
and then name some number like 0.0000000034.

The easier way is to notice that zero is less than any positive number.

Advanced algebra students will also notice that any negative number is always less than any positive number. If they were reading this book, they might have answered: -973.868.

490. The chances of that happening are $\dfrac{1}{100} \times \dfrac{1}{100}$ which is $\dfrac{1}{10,000}$
Rephrase the previous sentence in percents.

The chances of that happening are $1\% \times 1\%$ (0.01 × 0.01 = 0.0001), which is 0.01%. Women are rarely color blind.

Complete Solutions and Answers | 500–502

500. Looking at the results of the previous three questions, write the equation.

    We know Joe ran 14t feet.

    We know that the pelican ran 16t feet.

    We know that the 16t feet that the pelican ran is 100 feet more than the 14t feet than Joe ran.

$$16t = 100 + 14t$$

    Solving that equation is the easy part. Getting from the English to the equation is harder.

501. A serving of Puffy Buffy has 39% more cayenne than a one-pound burrito, which has 1.3 grams. How much cayenne does a serving of Puffy Buffy have?

    Long way: 39% of 1.3 g is 0.39 × 1.3 g, which is 0.507 g

        Then add 0.507 g to the original 1.3 g to get 1.807 g

    Short way: 39% plus the original 100% is 139%. (That step can always be done in your head.) Then 139% of 1.3 g is 1.39 × 1.3 g, which is 1.807 g of cayenne* in a serving of Puffy Buffy

502. When Jan got back to her apartment, she headed to her television set, which was 14 feet away. She was tired. She walked at 3 feet per second. Using fractions, find out how long she walked.

    This problem can be done in several ways.

First way: Using conversion factors, where 3 feet matches with one second. $\frac{14 \text{ feet}}{1} \times \frac{1 \text{ second}}{3 \text{ feet}} = \frac{14}{3} = 4\frac{2}{3}$ seconds

Second way: d = rt becomes $\qquad 14 = 3t$

    Divide both sides by 3 $\qquad\qquad \frac{14}{3} = t$

    Do the arithmetic $\qquad\qquad\qquad 4\frac{2}{3} = t$

---

* Cayenne (pronounced kai-N) is a red-powered spice that is hot. It bites. You can't believe everything on the Internet. I looked up `cayenne health` and found that it supposedly kills lung cancer cells, stops heart attacks, and keeps teeth from rotting. I put it on pizza just because it tastes good.

| 505–508 | Complete Solutions and Answers |

505. As they talked, she drank one-seventh of the cup and he drank one-fifth of the cup. What fraction of the cup was left?

Together they drank $\frac{1}{7} + \frac{1}{5}$ of the cup. $\frac{5}{35} + \frac{7}{35} = \frac{12}{35}$

What was left was $1 - \frac{12}{35} = \frac{35}{35} - \frac{12}{35} = \frac{23}{35}$

506. Balance $\qquad C_3H_8 + O_2 \rightarrow CO_2 + H_2O$

There are 3 C's on the left and 1 C on the right.
Add a couple of C's to the right side. $\qquad C_3H_8 + O_2 \rightarrow 3CO_2 + H_2O$

There are 8 H's on the left and 2 on the right.
Add 6 H's to the right side. $\qquad C_3H_8 + O_2 \rightarrow 3CO_2 + 4H_2O$

There are 2 O's on the left and 10 on the right.
Add 8 O's to the left side. $\qquad C_3H_8 + 5O_2 \rightarrow 3CO_2 + 4H_2O$

The C's are in balance (3 on each side).
The H's are in balance (8 on each side).
The O's are in balance (10 in each side). ☺

507. Janice and Junior had arrived there at 11:58 a.m. and it was now 12:32 p.m. It was time to leave. How long had they been visiting mom?

    From 11:58 to noon   2 minutes
    From noon to 12:32   32 minutes

There had visited with mom for 34 minutes.

508. There are two arteries coming out of Janice's heart. Name them.

The main one, the aorta, is the 120 pressure one that sends food and oxygen to muscles, brains, nerves, bones, etc.
The second one, the pulmonary artery, is the 15 pressure one that sends blood to her lungs so that it can get a breath of fresh air (get oxygenated). This was described in question #258.

Complete Solutions and Answers | 525–527

525. You and your best friend Jan head off to Ivy's Ice Cream. Ivy has 1,000 ice cream flavors.
You like 600 of Ivy's flavors. Jan like 550 of Ivy's flavors.
Let set K = the flavors you like.
Let set L = the flavors Jan likes.
Must the union of K and L (known as K ∪ L) equal all of Ivy's flavors?

No. For example, if you like the first 600 flavors on Ivy's list and Jan like the first 550 flavors on Ivy's list, then there would be many flavors that neither of you liked.

Must K and L be disjoint?

No. In fact, they couldn't be disjoint. If there are 1,000 flavors and you like 600 of them and Jan likes 550 of them, there have to be flavors that both of you like.

526. Jan slept from 10:07 a.m. to 5:45 p.m. How long had she slept?

From 10:07 to 11:00 is 53 minutes.
From 11:00 to noon is 1 hour.
From noon to 5:00 is 5 hours.
From 5:00 to 5:45 is 45 minutes.

```
                53 minutes
             1 hour
             5 hours
  +            45 minutes
             6 hours  98 minutes
```

98 minutes = 60 minutes + 38 minutes = 1 hour + 38 minutes

6 hours and 98 minutes = 7 hours and 38 minutes.

527. After two weeks the excitement over the aNchoVY-LaMB With a RiBBoN oF YaM song had died. The 198 ice cream orders per day had dropped by 97.5%. Approximately how many orders were now being received?

If it had dropped by 97.5%, then 2.5% remained.
100% − 97.5% = 2.5%
2.5% of 198 means 0.025 × 198, which is 4.95 which rounds to 5.
Ivy was now getting 5 orders per day.

| 528-530 | Complete Solutions and Answers |

528. In calculus, how will you write the expression for the area under the curve y = log x from x = 3 to x = 5?

$$\int_{x=3}^{5} \log x \, dx$$

When your parents see you writing this, they are going to faint—especially if you tell them it's easy to do. They will want to enroll you in genius school.

"Hey, Mom. You want to see the area under the curve $y = 6x^{5.4} - \sqrt{x-5}$ from x = e to x = π?"*

529. On which of these can we use cross multiplying?

    A) $\frac{4}{5} = \frac{x}{6}$ Yes. It is in the form $\frac{a}{b} = \frac{c}{d}$

    B) $\frac{2}{3} + \frac{4}{x} = \frac{7}{8}$ No. It is not in the form $\frac{a}{b} = \frac{c}{d}$

    C) $\frac{5}{x} = \frac{1}{2}$ Yes. It is in the form $\frac{a}{b} = \frac{c}{d}$

    D) $\frac{11x}{100} = \frac{5}{7}$ Yes. It is in the form $\frac{a}{b} = \frac{c}{d}$

530. Name some more absolute adjectives.

How about *chiefest*? You couldn't be more chiefest? Or *residual*? You couldn't be most residual.

Or *square* or *perpendicular* or *circular*? Either two lines are perpendicular (at right angles to each other) or they are not. My two lines can't be more perpendicular than your two lines.

How about *eternal*?

Can your task be more *impossible* than mine?

Can two numbers that you name be more *equal* than two numbers I will name?

---

*

Of course, it's just $\int_{x=e}^{\pi} (6x^{5.4} - \sqrt{x-5}) \, dx$

Complete Solutions and Answers 531–532

531. Which of these is/are not bones?

sternum, cranium, pinna, eyeball, femur, patella

The pinna (ear) is made of cartilage (not bone). The eyeball is not bone.

You may have noticed that skulls don't have eyes, noses, or ears—skulls are just bones (and teeth).

532. In the previous problem, saving $130 each month and earning 0.9% on that money each month will give you at the end of 72 months

$$130 \times \frac{(1 + 0.009)^{72} - 1}{0.009}$$

Give me the **general formula for an annuity** if you save d dollars for n months and are earning i interest each month.

At the end of n months you will have $d \times \frac{(1 + i)^n - 1}{i}$

That can be a handy formula if you want to do a little planning in your life. Let's take a real example (and I'll do the work).

Hey. I, your reader, like that. Do all the examples you like— while I watch!

One example will be enough.

Suppose you're making $24/hour. That's $48,000 a year ($24 per hour times 40 hours per week times 50 weeks per year).

If you can save one-tenth of that—which is not really that hard to do unless you are **spend crazy**. That's $4,800 saved per year. That's $400 per month.

Do that from age 22 to age 42—twenty years and earn six percent per year, which is 0.5% per month.

How much will you have at age 42? $400 \times \frac{(1 + 0.005)^{240} - 1}{0.005}$

The x^y key on my scientific calculator tells me that $(1.005)^{240} \doteq 3.310$

$$\doteq 400 \times \frac{3.310 - 1}{0.005}$$

$$= \$184{,}800.$$

Most 42-year-olds in this country don't have $184,800 in savings. You will.

| 541–542 | **Complete Solutions and Answers**

541. In percentage terms how much bigger in area is Stanthony's 20-inch pizza than his 16-inch pizza? (Again, use 3 for π.)

Solving this problem takes more than one step.
In the previous problem we found that the area of the 16-inch pizza is approximately 192 square inches.
The area of a 20-inch pizza is approximately 300 square inches.
The 20-inch pizza is 108 square inches bigger than the 16-inch pizza. (300 – 192 = 108)
Translation: In going from 192 to 300 we gained 108 square inches.
108 is what percent of 192?
108 is ?% of 192.
We don't know both sides of the *of* so we divide the number closest to the *of* into the other number.
108 ÷ 192 = 0.5625 = 56.25%
The 20-inch pizza is 56.25% larger than the 16-inch.

542. "If you sold your 16 cats for $4 each, think of all the money you could make." (Do this problem in your head without writing anything down.)

Learning to do a bit of mental arithmetic is good training.

$$\begin{array}{r} \overset{2}{1}6 \\ \times\ 4 \\ \hline 64 \end{array}$$

The more you do, the better you get at it. Sometimes in the morning when my brain is awake but my eyes haven't popped open yet, I play with math things, such as proving the Pythagorean theorem (from geometry) or the Law of Cosines ($c^2 = a^2 + b^2 - 2ab \cos C$, from trig). Other times I think of what pizza I will have for breakfast.

When you become an adult, you could have a six-pound bar of chocolate for breakfast every day if you wanted to. But most adults don't do that. Why? Because most adults have grown up.

Wait! I, your reader, have a question. Do you have pizza for breakfast?

Don't tell anybody, but I often do. The truth is either: ① I am six years old, or ② I, like Mary Poppins, have never grown up.

Complete Solutions and Answers 549–555

549. Solve
$16t = 100 + 14t$
$2t = 100$ Subtract 14t from both sides.
$t = 50$ Divide both sides by 2.

Joe (and the pelican) both ran for 50 seconds.

552. Jan had 4 hours of free time. She spent 2 hours and 35 minutes shopping. How much time was left?

```
   4 hours                    3 hours   60 minutes
-  2 hours    35 minutes    - 2 hours   35 minutes
                              1 hours   25 minutes
```

Jan had 1 hour and 25 minutes.

555. Suppose Sally had asked for seven quadrillion dollars instead of just one quadrillion dollars. If she distributed that evenly to the seven billion people, how much would each receive?

Using the **General Rule** let's simplify the problem. If you distribute $12 to 3 people, each one will receive $4. We divided.

So we divide seven quadrillion by seven billion.

$$\frac{7,000,000,000,000,000}{7,000,000,000}$$

We are going to do this problem three different ways. They will all get the same answer.

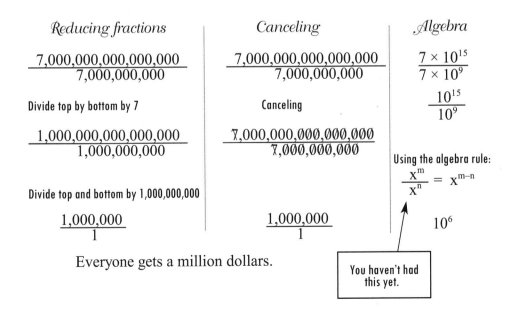

Everyone gets a million dollars.

183

557 Complete Solutions and Answers

557. Jan: "Three hundred pounds of fresh bear meat will feed us for a half year." Do Jan's numbers make sense—or is she exaggerating?

Wait a minute! I, your reader, am confused. This is supposed to be a math book. You are supposed to give me problems like $\frac{1}{4} + \frac{2}{5}$ or like six is what percent of nine.

This problem—"Do the numbers make sense?"—is more like real life. How am I supposed to handle that?

Maybe . . . this isn't a math book. Maybe this is a real-life book.

Someday you may be living a "real life"—a life in which the challenges/problems/situations are not always very neat. *Being a parent*, for example, can present almost unsolvable problems every day:

✸ Your kid is ten months old and won't stop crying.

✸ He is five years old and asks, "Why is there bad stuff in the world?"

✸ She is 15 and wants to get her nose tattooed pink because her friends are all doing that.

✸ He is 17 and is being chased by this 20-year-old who has everything wrong with her. He is saying, "But she loves me."

✸ She is 23 and wants to come back home to live with you. Five years of college "wasn't much fun." She will be happy to live in the basement and play video games indefinitely.

Okay. How do we do the bear meat problem?

I'm glad you asked.

300 pounds for the two of you means 150 pounds for each of you. Will 150 pounds feed you for a half year?

A year is about 300 days. A half year is about 150 days. That means about a pound of bear meat every day.*

That's M.T.E.B.M. (More Than Enough Bear Meat)

---

* There is no need in a problem like this to go into a lot of heavy-duty computation. $\frac{150 \text{ pounds}}{\frac{1}{2} \text{ year}} \times \frac{1 \text{ year}}{365\frac{1}{4} \text{ days}} =$ 0.8213552361396303901437371663244 pounds per day.

Complete Solutions and Answers | 558–559

558. That morning Jan had done 2½ hours of tap dancing. According to Prof. Eldwood's *Tap Dancing through Life*, 1843, a person spends 892 Calories in tap dancing for 40 minutes (40 minutes = ⅔ hour).

How many Calories had Jan spent tap dancing that morning?

The conversion factor is either $\dfrac{\text{⅔ hour}}{\text{892 Calories}}$ or $\dfrac{\text{892 Calories}}{\text{⅔ hour}}$

The top and bottom of these fractions are equal to each other.

We start with 2½ hours, and we want to convert that into Calories.

$$\dfrac{\text{2½ hours}}{1} \times \dfrac{\text{892 Calories}}{\text{⅔ hour}} = 3{,}345 \text{ Calories}$$

On your calculator—if you used one—you multiply 2.5 times 892 and then divide by 0.66666666666666666 and get 3345.0003.

Prof. Eldwood's statement that a person spends 892 Calories when tap dancing for 40 minutes is only an approximation. Much depends on (1) how hard you dance and (2) your body weight. Bigger people use more Calories because they have bigger bodies to move around.

It even depends on how cold the room is. Your body spends Calories just to stay warm.

559. A carton of Ivy's strawberry ice cream costs 42¢. If a cat owner buys 3 cartons of strawberry ice cream and then trades it for 7 cartons of aNChoVY-LaMB With a RiBBoN oF YaM ice cream, they have very cheap cat food that their cats adore. How much would a carton of this cat food cost the cat owner?

$$\dfrac{\text{1 carton anchovy}}{1} \times \dfrac{\text{3 strawberry}}{\text{7 anchovy}} \times \dfrac{42¢}{\text{one strawberry}}$$

$$= \dfrac{3 \times 42¢}{7} = 18¢$$

Eighteen cents for three cartons of cat food. That's a bargain. Back in problem #830, cartons of aNChoVY-LaMB With a RiBBoN oF YaM ice cream were much more expensive. That is because people had to buy that whole package of 6 cartons of ice cream and plastic spoon in order to get in and see Jan perform. They actually paid the $24.62 for three things: the ice cream, the spoon, and a "cover fee" to see her perform.

560–561 Complete Solutions and Answers

560. Convert $\frac{5}{6}$ to a percent.

The long way: First convert $\frac{5}{6}$ to a decimal

$$\frac{5}{6} = 0.83\frac{2}{6} = 0.83\frac{1}{3}$$

```
      0.83 R 2
   6) 5.00
      48
      ──
      20
      18
      ──
       2
```

Second, convert the decimal into a percent by moving the decimal point two places to the right. $0.83\frac{1}{3} = 83\frac{1}{3}\%$

The short way: $\frac{5}{6}$ was one of the eleven conversions that you memorized in *Life of Fred: Decimals and Percents*.

$\frac{1}{2} = 50\%$

$\frac{1}{3} = 33\frac{1}{3}\%$ $\frac{2}{3} = 66\frac{2}{3}\%$

$\frac{1}{4} = 25\%$ $\frac{3}{4} = 75\%$

$\frac{1}{8} = 12\frac{1}{2}\%$ $\frac{3}{8} = 37\frac{1}{2}\%$ $\frac{5}{8} = 62\frac{1}{2}\%$ $\frac{7}{8} = 87\frac{1}{2}\%$

$\frac{1}{6} = 16\frac{2}{3}\%$ $\frac{5}{6} = 83\frac{1}{3}\%$

561. Ivy's annual income this year is $2,500,000. She anticipates that it will grow by 4% per year in the future. How much will her income be two years from now?

This year	$2,500,000
Next year	$2,500,000 × 1.04 = $2,600,000
Two years from now	$2,500,000 × 1.04 × 1.04 = $2,704,000

What is fun is if you (or your older sister) have a scientific calculator. That's the kind that has sin, cos, tan, log and y^x keys on it.

The y^x key is sometimes it is labeled as the x^y key.

Suppose I want to find Ivy's annual income 20 years from now. That would be $2,500,000 × $(1.04)^{20}$. To do that on your basic calculator would require $2,500,000 × 1.04. Yuck. No thank you.

Using the y^x (and the owner's manual for your calculator) you can easily compute $(1.04)^{20}$. It's equal to 2.191123.

So $2,500,000 × $(1.04)^{20}$ would be equal to $5,477,807. That would be a very nice annual income.

Complete Solutions and Answers | 562–565

562. Convert 444 seconds into minutes and seconds.

```
       7 R 24
   60) 444
     - 420
       24
```

444 seconds equals 7 minutes and 24 seconds

563. The best way to combine together $+87 - 100 + 12 - 120 - 500$ is to combine together the positive numbers. Then combine together the negative numbers. Then combine the one positive number with the one negative number. Do it.

```
   87           -100
   12           -120
  +99           -500
                -720
```

And now combine $+99 - 720$. A gain of 99 and a loss of 720 equals $-\$621$.

When you are acting, don't wave at friends in the audience, forget what you lines are, and don't hit other actors.

564. Set A = {a gray cat, a quart of Ralph's litter, a scoop}
Set B = {a black cat, a quart of Ralph's litter, one leash}
Set C = {a gray cat, a cat toy, a collar}
Set D = {a black cat, a scoop, a cat toy}
Which two sets are disjoint?

Sets B and C have nothing in common. They are disjoint.

565.
$$(3\tfrac{1}{3})^3 = 3\tfrac{1}{3} \times 3\tfrac{1}{3} \times 3\tfrac{1}{3}$$
$$= \tfrac{10}{3} \times \tfrac{10}{3} \times \tfrac{10}{3}$$
$$= \tfrac{1000}{27}$$
$$= 37\tfrac{1}{27}$$

Notice that if you just multiply the 3s together ($3 \times 3 \times 3 = 27$) and multiply the $\tfrac{1}{3} \times \tfrac{1}{3} \times \tfrac{1}{3}$ together ($= 1/27$), your final answer isn't right.

In algebra, we will write $(3 + \tfrac{1}{3})^3 \neq 3^3 + (\tfrac{1}{3})^3$

> 575–577

Complete Solutions and Answers

575. You (not Cassie) have long eyelashes. You have 16 great-great grandparents. Is it possible that only one of those 16 had long eyelashes?

Yes. Your great-great grandma Lily was SL (and had long eyelashes). All the other 15 were SS.

Lily's son, Fitsell, was SL. Fitsell was one of your great grandparents.

Fitsell's daughter, Quash, was SL. You called her grandma Quash. Quash's daughter, Minder, was your mom and she was SL.

And you received an L from her and had long eyelashes.

Except for Lily, Fitsell, Quash, and Minder, all the rest were SS and had short eyelashes.

15 in Lily's generation out of 16 had short eyelashes
7 in Fitsell's generation out of 8 ———
3 in Quash's generation out of 4 ———
1 of your parents ———
. . . all these 26 were SS.

576. (continuing the previous problem) How long did they drive before they got to the parking lot of Beth's Vet? Give you answer in minutes and then convert that to hours.

From the previous problem we know that they were in the car for 21 minutes. Five of those minutes were spent in the parking lot. They spent 16 minutes (21 − 5 = 16) driving to the hospital.

Using a conversion factor,

$$\frac{16 \text{ minutes}}{1} \times \frac{1 \text{ hour}}{60 \text{ minutes}} = \frac{16}{60} = \frac{4}{15} \text{ hours.}$$

577. The set of Jan's favorite activities = {dancing, singing, acting, shopping, napping}.

The set of Joe's favorite activities = {fishing, eating, napping}. Are these sets disjoint?

No. Two sets are disjoint if there aren't things that belong to both sets. In this case, napping belongs to both sets.

Complete Solutions and Answers | 578–580

578. In the previous chapter (in the answer to problem #424) we learned that log 10^3 is 3 and log 10^7 is 7 and log 10^π is π. If you have a scientific calculator, you can easily find log 4 ≈ 0.60206.
Now I want you to tell me what $10^{0.60206}$ is approximately equal to.

It's just a matter of putting the pieces together.
Piece #1: You know that log 10^3 is 3 and log 10^7 is 7 and log 10^π is π—so log $10^{0.60206}$ = 0.60206.
Piece #2: You are given: log 4 ≈ 0.60206.

Let me write those pieces right next to each other.
$$\log 10^{0.60206} = 0.60206$$
$$\log 4 \approx 0.60206$$

So it would be a good guess that $10^{0.60206}$ would be equal to 4.
If you think that you have enough brains to go to college, then you will probably understand this by the time you are 18.
If you understand this right now, you shouldn't go to college—you should join Fred and teach in college.

579.
	5x + 0.6 + 7x = 39
Combine like terms	12x + 0.6 = 39
Subtract 0.6 from both sides	12x = 38.4
Divide both sides by 12	x = 3.2

580. Six buckets of Ralph's eyelashes plus a $2.98 tube of eye glue cost $248.98. How much does a bucket of Ralph's eyelashes cost?

Let x = the cost of one bucket of Ralph's eyelashes
Then 6x = the cost of six buckets.
Then 6x + 2.98 = the total cost.

The equation becomes	6x + 2.98 = 248.98
Subtract 2.98 from both sides	6x = 246
Divide both sides by 6	x = 41

A bucket of Ralph's eyelashes costs $41.

| 582–584 | Complete Solutions and Answers

582. They looked at a map of the lake. The map said that 3.4 cm (centimeters) corresponds to 5 miles. With a ruler they measured the distance around the lake. It was about 4 cm. To the nearest tenth of a mile, how far is it around the lake?

There are two ways to do this problem.

First way: Use a conversion factor. We want to convert 4 cm on the map into actual miles.

$$\frac{4 \text{ cm}}{1} \times \frac{5 \text{ miles}}{3.4 \text{ cm}} \approx 5.88235 \doteq 5.9 \text{ miles}$$

Second way: Set up a proportion. Let x = the number of miles that corresponds to 4 cm on the map.

$$\frac{3.4 \text{ cm}}{5 \text{ miles}} = \frac{4 \text{ cm}}{x \text{ miles}}$$

> This second way may be new to you. The technical term for that is **education**.

Cross multiply $3.4x = 20$

Divide both sides by 3.4 $x \approx 5.88235 \doteq 5.9$ miles

583. Half of her stuff was worthless and needed to be tossed out. A third of her stuff needed to be given to a thrift store. A tenth of her stuff was borrowed and needed to be returned to friends who had lent it to her. How much would she have left?

$$\frac{1}{2} + \frac{1}{3} + \frac{1}{10} = \frac{15}{30} + \frac{10}{30} + \frac{3}{30} = \frac{28}{30}$$

She would have left $1 - \frac{28}{30} = \frac{30}{30} - \frac{28}{30} = \frac{2}{30} = \frac{1}{15}$

584. What percent of men would be color blind?

We are supposing that one-hundredth (1%) of all the X chromosomes in the world have c.

Men have only one X chromosome. Ninety-nine percent of the time it will have C. One percent of the time it will have c, and the man will be color-blind.

1% of the men are color-blind.

0.01% of women are color-blind.

Assuming that 1% of all X chromosomes in the world have c, that would mean that men are 100 times more likely than women to be color-blind.

Complete Solutions and Answers | 591–593

591. Given the fact that 9 liters of fluid is being dumped into that twenty-foot tube every day, why don't you have to drink 2.4 gallons each day?

    Happily, after the **WETLANDS** of the small intestine is the great drying out place called the large intestine. In the large intestine much of the fluids from the wetlands are reabsorbed back into body.

    Otherwise, you would have constant diarrhea. (If you don't know what diarrhea means, you could either look that up in a dictionary, or you could do as I did and ask my mother. She explained it to me, "It's soupy poopy.")

592. When you got to Jan's apartment, you counted her cats. There were 21 of them, some big and some very small. Where did the other five come from?

    This is a biology book.   Cows have calves.
    Lions have cubs.
    People have babies.
    Cats have kittens.

Jan named her new 5 cats: Quincy, Rats, Slender, Tagalong, and Uranus.
    I, your reader, have a silly question.
    What is it?
    With five more cats, Jan is going get to the end of the alphabet. After she had Vettles, Wango, Xander, Youngtown, and Zenda **what is she going to call her next cat?**

    I asked Jan. She told me that she was going to resort to mathematics. (Math can be very useful.) She said that she will call her next cats: cat #27, cat #28, cat #29, cat #30, cat #31, cat #32, cat #33, cat #34, cat #35, cat #36, cat #37. . . .
    Thank you.

593. Humans normally have about 5 quarts of blood. You have lost 1.7 quarts. What class of hemorrhaging (bleeding) are you experiencing?

    1.7 quarts is what percent of 5 quarts?
    1.7 = ?% of 5
    $1.7 \div 5 = 0.34 = 34\%$

    You are in Class 3: 30 < blood loss < 40%. You have a legitimate reason to pass out. You did.

| 594–596 | Complete Solutions and Answers

594. While Jan was unconscious, the little girl had eaten some of Jan's chocolate doughnuts. Jan's stash of 6 pounds of doughnuts was now 3 ounces lighter. (1 pound = 16 ounces). How much was left?

```
      6 lbs.              5 lbs.   16 ozs.
 −           3 ozs.    −           3 ozs.
                          5 lbs.   13 ozs.
```

Jan had 5 pounds and 13 ounces of doughnuts left.

595. $6\frac{2}{5} \div 2\frac{2}{7}$

$\frac{32}{5} \div \frac{16}{7}$

$\frac{32}{5} \times \frac{7}{16}$

$\frac{\overset{2}{\cancel{32}}}{5} \times \frac{7}{\underset{1}{\cancel{16}}} = \frac{14}{5} = 2\frac{4}{5}$

596. You have the same number of nickels, dimes, and quarters in your pocket.
    Let x = the number of nickels you have.
A) How many dimes do you have?
    You have x dimes (since the number of nickels and the number of dimes are equal.)
B) How many quarters do you have?
    You have x quarters.
C) How many cents are your nickels worth?
    (If you had 7 nickels, they would be worth 7(5).
    If you had 8 nickels, they would be worth 8(5).
    You have x nickels.)
    Your x nickels are worth 5x cents.
D) How many cents are your dimes worth?
    Your x dimes are worth 10x cents.
E) How many cents are your quarters worth?
    Your x quarters are worth 25x.
F) How many cents are your nickels, dimes, and quarters worth?
    They are worth a total of 5x + 10x + 25x (or, if you like, 40x).

Complete Solutions and Answers | 600–603

600. $$CO + H_2 \rightarrow CH_3OH$$

The O's are in balance.
The C's are in balance.
There are 2 H's on the left and 4 H's on the right.

$$CO + 2H_2 \rightarrow CH_3OH$$

Hey! That was pleasant.
    The C's, the H's, and the O's are all in balance.

601. Twenty-five sparklettes weigh $9\frac{3}{8}$ pounds. How much did each one weigh?

To figure out whether to add, subtract, multiply, or divide, you can use the 𝕲𝖊𝖓𝖊𝖗𝖆𝖑 𝕽𝖚𝖑𝖊, which suggests that you use really simple numbers and see what operation you are using.

If 3 sparklettes weighed 12 pounds, then each one would weigh 4 pounds. You divided.

$$9\frac{3}{8} \div 25 = \frac{75}{8} \div 25 = \frac{75}{8} \times \frac{1}{25} = \frac{\overset{3}{\cancel{75}}}{8} \times \frac{1}{\underset{1}{\cancel{25}}} = \frac{3}{8}$$

Each sparklette weighed $\frac{3}{8}$ of a pound.

602. Jan had received her last paycheck—$93.44. *At least I won't starve,* she thought to herself. How many two-dollar pies could she buy at Ivy's?

Again, we can use the 𝕲𝖊𝖓𝖊𝖗𝖆𝖑 𝕽𝖚𝖑𝖊, which suggests that you use really simple numbers and see what operation you are using. If Jan had $12 and pies cost $2, she could buy 6 pies. That's division.

93.44 ÷ 2 = 46.72 pies.

Of course, you can't buy 0.72 pies. Jan could buy 46 pies and have a little money left over.

In this case we did not round 46.72 up to 47. If you can buy <u>at most</u> 46.72 pies, then you can't buy 47 pies.

603. Is 93.44 evenly divisible by 3 or by 5?

Since the digits add to 20 (= 9 + 3 + 4 + 4) and since 20 isn't divisible evenly by 3, neither is 93.44.

Since the last digit of 93.44 isn't either a 0 or a 5, 93.44 isn't evenly divisible by 5.

193

| 607–611 | Complete Solutions and Answers

607. Junior had owned a four-dollar bike and 26¢ in change. He now had 26¢. To the nearest percent, what percent of his "wealth" had he lost?

He had lost $4 of his original $4.26 "fortune."
$4 is what percent of $4.26?
4 = ?% of 4.26
4 ÷ 4.26 ≈ 0.93897 = 93.897% ≐ 94% of his wealth was gone.

611. How long, in weeks, is 5,000 hours of preparation?

I, your reader, object to this question. It seems really vague. You haven't told me enough information. Regular math books say something like . . .

Johnny wants to spend 5,000 hours chopping wood. He can work 6 hours each day. How many days will it take him to chop all that wood?

I, your author, always hated those "regular" math books. Who in blazes ever cared about Johnny chopping wood? And why couldn't Johnny just buy some wood that was pre-chopped?

When Janice and Junior are planning their life together, their success in J̶a̶n̶s̶'̶ ̶J̶a̶m̶s̶ depends greatly on their spending those 5,000 hours in preparation. In the United States, three out of four food businesses fail within two years. And most of those are because of ❶ lack of start-up capital and/or ❷ lack of preparation.

But why didn't you tell me how much time each day they would be spending learning about the jam business?

Because in REAL LIFE you aren't told. You decide.
If you spend 40 hours per week, then 5,000 hours is

$$\frac{5000 \text{ hours}}{1} \times \frac{1 \text{ week}}{40 \text{ hours}} = 125 \text{ weeks}$$

If you spend 8 hours per week, then 5,000 hours is

$$\frac{5000 \text{ hours}}{1} \times \frac{1 \text{ week}}{8 \text{ hours}} = 625 \text{ weeks}$$

625 weeks! That's about 12 years! Get real!

I am. Working hard, it's 125 weeks (≈ 2½ years). What do you think your late teenage years were invented for? Playing video games?

Complete Solutions and Answers | 624–625

624. (continuing the previous problem) If your mother has long eyelashes, what can you say about your grandfather's *phenotype*?

In the previous problem we found that your grandfather must be either SS or SL.

Since your mom has long eyelashes and since her mom (your grandmother) is SS, your mom must have received an L from her father.

Therefore, he must be SL.

He must have long eyelashes. That's his phenotype.

625. Darlene wasn't paying attention to her driving. She was thinking of Joe and how cute he was. She was doing 60 mph on a city street.

The cop was $\frac{2}{5}$ of a mile behind her when he turned on his lights. He was going 84 mph. How long before he caught up?

Let t = the number of seconds till he caught up with Darlene.
Then 60t = the number of miles Darlene went.
Then 84t = the number of miles that the cop went.
Since the number of miles that the cop went was $\frac{2}{5}$ of a mile more than the number of miles that Darlene went . . .

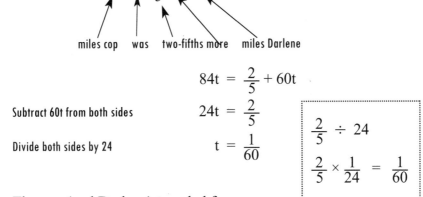

	$84t = \frac{2}{5} + 60t$
Subtract 60t from both sides	$24t = \frac{2}{5}$
Divide both sides by 24	$t = \frac{1}{60}$

$$\frac{2}{5} \div 24$$
$$\frac{2}{5} \times \frac{1}{24} = \frac{1}{60}$$

The cop (and Darlene) traveled for one-sixtieth of an hour, which is one minute.

| 626–629 | **Complete Solutions and Answers**

626. At the cells, oxygen is dropped off. Where in the circulatory system was the oxygen picked up?

You may have noticed that you are in the habit of breathing: sucking in air and blowing it out. There is a reason for this. When you take in air into your lungs, oxygen moves from the air into your blood stream. Also carbon dioxide (CO_2) goes from your blood stream through your lungs back into the air you expel.

Your grocery store for cells: your small intestine.
Your gas station for your cells: your lungs.

627. Is the set of herbivores a subset of the set of omnivores?

No. All I have to do to show that the set of herbivores is not a subset of omnivores is to find one herbivore that isn't an omnivore. That's easy: cows. Recall: A herbivore *only* eats plants.

In fact, it is super easy since every herbivore isn't an omnivore. The two sets are disjoint.

628. How many years and weeks is 1,650 weeks? (Assume a year contains 52 weeks.)

To convert 135 minutes into hours and minutes: $60\overline{)135}$ 2 R15
135 minutes = 2 hours and 15 minutes.

To convert 317 inches into feet and inches: $12\overline{)317}$ 26 R 5
317 inches = 26 feet and 5 inches.

To convert 1650 weeks into years and weeks: $52\overline{)1650}$ 31 R 38
1650 weeks = 31 years and 38 weeks.

629. What is the cardinal number associated with {Jan, Juno, Charlie}?

The cardinal number of a set is the number of members in that set. The cardinal number of {Jan, Juno, Charlie} is 3.

The cardinal number of { } is 0.

The cardinal number of your ankles that haven't been chewed on is one.

Complete Solutions and Answers 639–641

639. The nine doughnuts and the 284 Calories of Sluice were 6,800 Calories. How many Calories were in each doughnut?

> We always begin the solution of a word problem with . . .
> Let x = the number of Calories in a doughnut.
> Then 9x = the number of Calories in the nine doughnuts she ate.
> Then 9x + 284 = the total Calories she ate.

After you write the Let x = . . . and these Then . . . statements, the equation almost writes itself.

$$9x + 284 = 6{,}800$$

Subtract 284 from each side $\quad 9x = 6{,}516$
Divide both sides by 9 $\quad x = 724$

Each doughnut was 724 Calories.

640. Jan looked at the clock. It was 4:40. She had to be at the theater by 6:10. How much time did she have?

> From 4:40 to 5:00 is 20 minutes.
> From 5:00 to 6:00 is 1 hour.
> From 6:00 to 6:10 is 10 minutes.

Jan had one hour and thirty minutes.

641. "I will take 21 glamour pictures of you and toss in a 79¢ frame. The whole thing will be $11." How much would each picture be?

> I'm going to work in dollars.
> Let x = the cost of one picture.
> Then 21x = the cost of all the pictures.
> Then 21x + $0.79 = the cost of the whole thing.

At this point the equation is easy to write. $\quad 21x + 0.79 = 11$

Subtract 0.79 from both sides. $\quad 21x = 10.29$
Divide both sides by 21 $\quad x = 0.49$

Each picture would cost $0.49, which is 49¢.

Please get in the habit of writing the "Let x = . . ." and the "Then . . ." statements rather than jumping from the English to the equations. It isn't that much work. When you get to algebra, the word problems will be much more varied. You'll have coin problems, distance-rate-time problems, age problems, and mixture problems. Leaping directly from English to equation will often result in disaster.

| 645–648 | **Complete Solutions and Answers** |

645. The cone had a radius of 3 inches and a height of 4 inches. To the nearest cubic inch, what was its volume?

Use 3.1 for π. $V_{cone} = (\frac{1}{3})\pi r^2 h$

$V_{cone} = (\frac{1}{3})\pi r^2 h$ becomes $V = (\frac{1}{3})(3.1)3^2(4)$
Doing the arithmetic $V = 37.2$
Rounding $V \doteq 37$

The volume of Ralph's cone-of-lashes is approximately 37 cubic inches.

646. If it was 5.9 miles around the lake and if the lake were circular, how far is it across the lake? Use 3 for π in this problem. Round your answer to the nearest tenth of a mile.

Circumference equals pi times diameter.
$C = \pi d$

Replace C by 5.9 and π by 3 $5.9 = 3d$
Divide both sides by 3 $1.96666666 \approx d$
Round to the nearest tenth $2 \doteq d$

It's roughly two miles across the lake.

648.   *If you haven't used it in a year,*
              seriously consider getting rid of it.

   Your question: Can you think of any exceptions to that rule?

   Exceptions are hard to find. Your list may be different than mine.
1. fire extinguishers
2. photo albums
3. some books
   I'm having trouble thinking of anything more to add to this list.

Complete Solutions and Answers

652. With 1,000 flavors, Ivy's menu was the size of a book. If she listed 23 flavors on each page, how many pages would her menu be?

Using the 𝔊𝔢𝔫𝔢𝔯𝔞𝔩 𝔑𝔲𝔩𝔢*, we suppose that there were 12 flavors and 3 flavors on each page. Then we would need 4 pages. We divided.

So in this problem we need to divide 1,000 by 23.

$1{,}000 \div 23 \approx 43.47$ pages.

Ivy's menu would need to have at least 44 pages.

Stop! I, your reader, notice that you have made an error.

Oh? I do that all the time. Where did I goof?

When you round off 43.47 to the nearest page you will get 43 pages, not 44 pages.

You are right, except for one small fact. Rounding is not the appropriate thing to do here. I know that Ivy will need *at least* 43.47 pages in order to list all 1,000 flavors. If Ivy only uses 43 pages, she won't be able to list all the flavors. 43 × 23 equals only 989 flavors.

653. Faster humans can run about 16 mph for a hundred yards. A bear can run about 25 mph over a distance of two miles. What percent faster is a bear than a human? (Round your answer to the nearest percent.)

Bears can run 9 mph faster than humans. (25 − 16 = 9)

9 is what percent more than 16?

9 = ?% of 16

$9 \div 16 = \dfrac{9}{16} = 0.5625 = 56.25\% \doteq 56\%$

Bears can run about 56% faster than humans.

If you meet a bear, one of the worst things you can do is run away. That gets them all excited. They think you want to play tag. The loser in this game of tag will not be the bear.

If you want to get to Heaven in a hurry, slap a 550-pound *Ursus acrtos* (brown bear) on his nose and yell, "You're it!"

You capitalize and italicize genera (*Ursus* = bear) and you italicize species (*acrtos* = brown bear).

---

* If you don't know whether to add, subtract, multiply or divide, first restate the problem with really simple numbers.

| 654–657 | Complete Solutions and Answers

654. Find the LCD for $\frac{5}{6}$ and $\frac{2}{15}$

One way to find the least common denominator of is to list the multiples of 6, which are 6, 12, 18, 24, 30, 36, 42 . . . and pick the first one that 15 divides evenly into.

The LCD is 30.

655. "At 6 p.m." It's now 4:15 p.m. How long does Jan have to get there?

From 4:15 to 5:00 is 45 minutes.
From 5 to 6 is one hour.
She had one hour and forty-five minutes to get to the rehersal.

656. One of Ivy's most beautiful ice creams was a golden yellow. It was a delight to the eye. *Zea mays* sold $2\frac{1}{4}$ gallons last month. How much more was that than the $\frac{7}{10}$ of a gallon that Sneaky Dog sold?

$$2\frac{1}{4} \qquad 2\frac{5}{20} \qquad 1\frac{20}{20} + \frac{5}{20} \qquad 1\frac{25}{20}$$
$$-\frac{7}{10} \qquad \frac{14}{20} \qquad \frac{14}{20} \qquad \frac{14}{20}$$
$$\qquad\qquad\qquad\qquad\qquad\qquad\qquad\qquad 1\frac{11}{20}$$

Zea mays sold $1\frac{11}{20}$ more gallons than Sneaky Dog.

657. $\frac{5}{6} + \frac{2}{15}$

$= \frac{5 \times 5}{6 \times 5} + \frac{2 \times 2}{15 \times 2}$

$= \frac{25}{30} + \frac{4}{30}$

$= \frac{29}{30}$

Complete Solutions and Answers | 658–661

658. 3 is what percent of 4?

    3 is ?% of 4

    When we don't know both sides of the *of* we divide the number closest to the *of* into the other number.

$$4\overline{)3.00} = 0.75 \quad 0.75 = 75\%$$

    3 is 75% of 4.

659. Ivy explained that she researched every ice cream store in the world and used every flavor she found.

    Does this mean that the list of flavors that Ralph's Three Flavor Ice Cream store and the list of flavors of Ivy's Ice Cream are not disjoint?

    They are not disjoint. Every flavor that Ralph has is a flavor that Ivy will have. All three of Ralph's flavors can be found on Ivy's list.

660. You think: *$200 per hour, 40 hours per week, 50 weeks per year.* What would be Jan's annual income?

$$\frac{\$200}{1 \text{ hour}} \times \frac{40 \text{ hours}}{1 \text{ week}} \times \frac{50 \text{ weeks}}{1 \text{ year}} = \$400{,}000 \text{ per year}$$

In general if you are working 40 hours per week for 50 weeks each year, that makes 2,000 work-hours per year. That makes it easy to convert any hourly wage into an annual wage.

    $10 per hour means $20,000 per year.   (10 × 2000 = 20,000)

    $40 per hour means $80,000 per year.

If someone makes $100,000 a year, that means $50 per hour.

    (100,000 ÷ 2,000 = 50)

661. After the show, Jan was giving hoofprint autographs to all her fans. Four prints plus a $5.23 album to put them in were on sale for $16.35. How much was each hoofprint autograph worth?

    Let x = the price of one print.
    Then 4x = the price of four prints.
    Then 4x + 5.23 = the total price.

    The equation becomes   4x + 5.23 = 16.35
   Subtract 5.23 from both sides   4x = 11.12
   Divide both sides by 4   x = 2.78

Each hoofprint autograph was worth $2.78.

| 662–664 | Complete Solutions and Answers

662. Ivy's annual expenses are:
Building and utilities $42,000.
Cost of ice cream $377,000.
Paying Jan to sing and dance $400,000.

    total expenses $819,000.

Ivy's annual income is:
$2,500,000.
What is Ivy's annual profit?

Profit equals your income minus your expenses.

$P = I - E$ becomes $P = 2,500,000 - 819,000$
$P = \$1,681,000$

663. Ivy's giant ice cream cone (see two problems ago) was shipped to the front of her store in a giant cylinder. You know from the current chapter that $V_{cone} = \frac{1}{3}\pi r^2 h$. Make a guess . . . the cone occupies what fraction of the volume of the cylinder that encloses it?

Any guess between ¼ and ½ would be a good guess.
The cone is 50' tall and 20' across. Its radius is 10'.
The cylinder is also 50' across, and its radius is also 10'.
$V_{cone} = \frac{1}{3}\pi r^2 h$
What's the volume formula for a cylinder? We had that in *Zillions of Practice Problems Pre-Algebra 0 with Physics*. $V_{cylinder} = \pi r^2 h$.
Looking at those two formulas, we can tell that the volume of the cone is *exactly* one-third of the volume of the cylinder that encloses it.
To me, that is surprising. There is no way to tell that by just looking at the picture. (In calculus we will be able to prove that $V_{cone} = \frac{1}{3}\pi r^2 h$.)

664. Jan: "The world is divided into two groups—theater arts majors and the audience."

✱ Are those two sets disjoint?
No. Sometimes Jan sits in the audience. That's an essential part of learning to be an actor. You need to see the stage as your audience sees it.

✱ Is the union of those two sets equal to the set of all the people in the world?
No. There are people who are neither actors nor members of the audience.

Complete Solutions and Answers | 665–669

665. She had 63 boxes of macaroni-and-cheese. Eating 3 boxes per day, how long would that last? For the practice, please use a conversion factor.

The conversion factor will be either $\dfrac{3 \text{ boxes}}{1 \text{ day}}$ or $\dfrac{1 \text{ day}}{3 \text{ boxes}}$

$$\dfrac{63 \text{ boxes}}{1} \times \dfrac{1 \text{ day}}{3 \text{ boxes}} = 21 \text{ days}$$

666. They charged you $5.24 for each bandage. They applied 19 bandages. How much were you billed for those bandages?

$\$5.24 \times 19 = \$99.56$

667. $\dfrac{5}{6} \times \dfrac{2}{15}$

$= \dfrac{\cancel{5}^1}{\cancel{6}_3} \times \dfrac{\cancel{2}^1}{\cancel{15}_3} = \dfrac{1}{9}$

668. Two is a nice number to start the Human Family. What do we know about their eyelashes?

First of all, they both couldn't have had short eyelashes. (Otherwise, all their descendants would have short eyelashes.)

Second, they both couldn't have phenotypes of LL, because then we would all have long eyelashes, which isn't true.

What is left are three possibilities:

    Case 1: One was LL and the other LS.
    Case 2: Both were LS.
    Case 3: One was LL and the other was SS.

669. (continuing the previous problems) How far did Darlene drive to get to the hospital?

We know she was going 12 mph (from problem #465). We know she drove $\dfrac{4}{15}$ hours (from problem #583).

$d = rt$ becomes $\quad d = \dfrac{12}{1} \times \dfrac{4}{15} = \dfrac{\cancel{12}^4}{1} \times \dfrac{4}{\cancel{15}_5} = \dfrac{16}{5} = 3\dfrac{1}{5}$ miles

| 671–678 | Complete Solutions and Answers

671. What's seven divided by one-millionth?

$$7 \div \frac{1}{1,000,000}$$

$$= \frac{7}{1} \times \frac{1,000,000}{1}$$

$$= 7,000,000$$

When you divide by a really small number, you get a really big answer. Dividing by a really small number is like asking how many ants could you fit in a automobile. Ants are small. You can get a zillion of them in a car.

676. To the nearest percent, what fraction of the cup was left?

Expressing $\frac{23}{35}$ as a percent, we first convert it into a decimal.

You use you calculator; I'm going to do it the old-fashioned way.

$$\begin{array}{r} 0.6571 \\ 35 \overline{)23.0000} \\ \underline{21\,0} \\ 200 \\ \underline{175} \\ 250 \\ \underline{245} \\ 50 \\ \underline{35} \end{array}$$

$\frac{23}{35} \approx 0.6571 = 65.71\% \doteq 66\%$

678. Is "third" a cardinal number?

Cardinal numbers are numbers used to count the members of a set. They are numbers like 0, 1, 2, or 3.

Third is an ordinal number.

Complete Solutions and Answers | 680–682

680. The speeding ticket was for $500. The speed limit was 35 mph. The fine was $x for each mph over the speed limit plus court costs of $75. Find the value of x.

> Darlene was 25 mph over the speed limit. (60 − 35 = 25)
> The fine was 25x + 75.
> The fine was $500.

The equation is easy after you have written out the English.

$$25x + 75 = 500$$

Subtract 75 from both sides $\quad 25x = 425$
Divide both sides by 25 $\quad x = 17$

The fine was $17 for each mph over the speed limit (plus $75 in court costs).

681. When Jan was memorizing lines for a play, she could learn 6 lines in 8 minutes. At that rate how many lines could she memorize in 44 minutes. Use a conversion factor.

Six lines matches up with 8 minutes. The conversion factor will be either $\dfrac{6 \text{ lines}}{8 \text{ min}}$ or $\dfrac{8 \text{ min}}{6 \text{ lines}}$

$$\dfrac{44 \text{ min}}{1} \times \dfrac{6 \text{ lines}}{8 \text{ min}} = \dfrac{44 \cancel{\text{ min}}}{1} \times \dfrac{6 \text{ lines}}{8 \cancel{\text{ min}}} = 33 \text{ lines}$$

682. At the dinner table if your mom asks you to pass her 180 grams of glucose, how many ounces of sugar would that be? (One gram is about 0.035 ounces.)   ←conversion factor

$$\dfrac{180 \text{ g}}{1} \times \dfrac{0.035 \text{ oz.}}{1 \text{ g}} = 6.3 \text{ ounces of sugar}$$

It is much faster to weigh out 6.3 ounces of sugar than to count 602,213,670,000,000,000,000,000 molecules of sugar. That might take all day. (understatement)

For fun, how long would that really be? Counting one molecule per second,

$$\dfrac{602{,}213{,}670{,}000{,}000{,}000{,}000{,}000 \text{ molecules}}{1} \times \dfrac{1 \text{ sec}}{1 \text{ molecule}} \times \dfrac{1 \text{ minute}}{60 \text{ sec}}$$

$$\times \dfrac{1 \text{ hour}}{60 \text{ min}} \times \dfrac{1 \text{ day}}{24 \text{ hrs}} \times \dfrac{1 \text{ year}}{365¼ \text{ days}} \doteq 19{,}082{,}999{,}657{,}768{,}651 \text{ years.}$$

| 687–690 | Complete Solutions and Answers

687. A mom came into Ivy's Ice Cream with her four kids. Here are the sets of flavors that each kid ordered:
Kid #1 = {chocolate, Shivering Strawberry}
Kid #2 = {Vacant Vanilla, Crunchy Cinnamon}
Kid #3 = {Shivering Strawberry, Crunchy Cinnamon}
Kid #4 = {Vacant Vanilla, chocolate, Crunchy Cinnamon}
Is any pair of these sets disjoint?

I first noticed that sets #2, #3, and #4 all have Crunchy Cinnamon in them, so no pair of these will be disjoint.
Next, I looked at whether #1 is disjoint from any of the rest.
#1 is disjoint from #2.
#1 is not disjoint from #3 (since they both have Shivering Strawberry).
#1 is not disjoint from #4 (since they both have chocolate).

690. Ear plugs—.10¢ a pair. How many pairs could you buy for a dollar?
.10¢ is not the same as 10¢
.10¢ is one-tenth of one cent.
How many .10¢ are there in 100¢?
Using the General Rule we substitute some simple numbers to see whether we add, subtract, multiply, or divide. Suppose we wanted to find out how many nickels (5¢) are in a quarter (25¢). There are five nickels in a quarter. We divided.

100¢ ÷ 0.10 = 1,000

You could buy a thousand pairs of ear plugs.

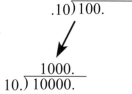

206

Complete Solutions and Answers | 700–702

700. Jan was embarrassed. She offered the movie star five apologies (all of equal length) and then said goodbye, which took 4 more seconds. Her 5 apologies and her goodbye took a total of 35 seconds. How many seconds did each apology take?

Let t = the number of seconds for each of Jan's apologies.
Then 5t = the time it took to make all five apologies.
Then 5t + 4 = the time Jan spoke.

The equation almost writes itself at this point. $5t + 4 = 35$
  Subtract 4 from both sides $5t = 31$
  Divide both sides by 5 $t = 6.2$

Each apology took 6.2 seconds.

701. If her 25 sparklettes cost a total of $4.25, how much did each one cost? Give your answer in cents.

$4.25 is 425¢

Using the **General Rule**, if 3 sparklettes cost 12¢, the each one would have cost 4¢. You divided.

$425 \div 25 = 17$. Each sparklette cost 17¢.

702. Potassium chlorate decomposes into potassium chloride and oxygen.

  Balance $KClO_3 \rightarrow KCl + O_2$

There are 3 O's on the left and 2 O's on the right.
Let's make 6 O's on each side. $2KClO_3 \rightarrow KCl + 3O_2$

There ae 2 K's on the left and 1 K on the right.
Let's add a K to the right side. $2KClO_3 \rightarrow 2KCl + 3O_2$

The K's are in balance (2 on each side).
The Cl's are in balance (2 on each side).
The O's are in balance (6 on each side). ☺

| 711-715 | **Complete Solutions and Answers**

711. How many pennies would you have to subtract from $93.44 in order to get a number that was evenly divisible by both 3 and 5?

In order to be evenly divisible by 5, the last digit must be either 0 or 5. So the possible candidates are . . .

93.40
93.35
93.30
93.25
etc.

We'll test each of these to see if it is divisible by 3.

93.40   The sum of the digits is 16. Since 16 isn't divisible by 3, neither is 93.40

93.35   The sum of the digits is 20. Since 20 isn't divisible by 3, neither is 93.40

93.30   The sum of the digits is 15. Since 15 is divisible by 3, so it 93.30.

You would have to subtract fourteen pennies. ($93.44 – $93.30)

714. Balance $B_2H_6 + O_2 \rightarrow B_2O_3 + H_2O$

The B's are in balance. 2B's on the left and 2 B's on the right.
Let's balance the H's. $B_2H_6 + O_2 \rightarrow B_2O_3 + 3H_2O$

Now balance the O's. $B_2H_6 + 3O_2 \rightarrow B_2O_3 + 3H_2O$

The B's balance (2 on each side). The H's balance (6 on each side). The O's balance (6 on each side). ☺

715. If an Ivy pie weighs 1.6 pounds, how much would a mole of those pies weigh?

A dozen is 12.
A score is 20.
A gross is 144.
A mole is 6.0221236×10^{23}.

$1.6 \times 6.0221236 \times 10^{23} = 9.63539776 \times 10^{23}$ pounds
or, if you like, 963,539,776,000,000,000,000,000 pounds.

Complete Solutions and Answers | 719–722

719. She had $91.02. He had $0.26. What percent more did she have than he had?

    She had $90.76 more than he had. (91.02 – 0.26 = 90.76)
    90.76 is what percent of 0.26?
    90.76 = ?% of 0.26

    90.76 ÷ 0.26 ≈ 349.077 = 34907.7% ≐ 34,908%

She had a lot more money than he had—if you are looking at it from a percent point of view. If you are looking at it from a dollars-and-cents point of view, these guys' combined fortunes were less than a hundred dollars. A dinner and a movie (with two orders of popcorn) and they would be broke.

721. She walked east at 4 feet per second. He walked west at 3 feet per second. In t seconds they were 4,620 feet apart. How long were they walking?

    Let t = the number of seconds each one walked. (The "let t =" is always the first thing you write down in solving algebra word problems.)
    Then 4t = the number of feet she walked. (If she walked for t seconds at the rate of 4 feet per second, the d=rt formula says that she walked for 4t feet.)
    Then 3t = the number of feet that he walked.
    Then 4t + 3t = the number of feet they both walked. They would be 4t + 3t feet apart since they were walking in opposite directions.

Now we can write the equation	4t + 3t = 4620
Combine like terms	7t = 4620
Divide both sides by 7	t = 660

They walked for 660 seconds.

722. Alfredo von Futz bought a purple house for $360,000. After he moved into it, it became a "house of a famous movie star." It was now worth $500,000. To the nearest percent what percent increase was that?

    The price increased by $140,000. (500,000 – 360,000)
    $140,000 is what percent of $360,000?
    140,000 = ?% of 360,000

If you don't know both sides of the *of*, you divide the number closest to the *of* into the other number.

    140,000 ÷ 360,000 ≈ 0.38888 = 38.888% ≐ 39%

| 723–725 | **Complete Solutions and Answers**

723. Ivy normally gets her hair done for $30. After hearing about Ivy's success, her hairdresser increased Ivy's bill to $35. What percentage increase was that?

$35 is an *increase* of $5 over $30.
5 is what percent of 30?
5 is ?% of 30

We don't know both sides of the *of* so we divide the number closest to the *of* into the other number.

```
        0.16 R20
     30) 5.00
         30
         200
         180
          20
```

$0.16 \frac{20}{30} = 0.16\frac{2}{3} = 16\frac{2}{3}\%$

Ivy's hairdresser increased her rate by 16⅔%.

724. A full 5-gallon container of **SHIVERING STRAWBERRY** weighs 37 pounds. Joe put the whole thing on a cone. It was the biggest "one scoop" that the world had ever seen.

70% of it melted and ran down his arm before Joe could finish it. How many pounds of **SHIVERING STRAWBERRY** did he eat?

If 70% melted, then 30% remained.
30% of 37 pounds
0.3 × 37
11.1 pounds he ate. He began to shiver.

725. $(4\frac{3}{7})^3 = 4\frac{3}{7} \times 4\frac{3}{7} \times 4\frac{3}{7}$

$= \frac{31}{7} \times \frac{31}{7} \times \frac{31}{7}$ To multiply mixed numbers you have to change them into improper fractions.

$= \frac{29791}{343} = 86\frac{293}{343}$

Even though you are working in fractions, calculators can come in handy.

Complete Solutions and Answers | 726–728

726. Exercise makes a body grow stronger. You can do 6 pushups. Jan can do 100% more pushups than you. How many can Jan do?

This is a "15% more than" problem. There are two ways to do it.
One way: 100% of 6 is 6.
Add 6 to the original 6 and get 12 pushups.

Another way: 100% more than means 200% (100% + 100%) of the original amount. 200% of 6 is 12 pushups.

One hundred percent more than something means twice as much.
Two hundred percent more than sometimes means three times as much.
Eight hundred percent more means nine times as much.

727. In the last year only 4 customers had ordered anchovy-lamb with a ribbon of yam. Today, with Jan's singing and dancing, 7 people got up the courage to order that flavor. Seven is what percent more than 4?

7 is an *increase* of 3 over the original 4.
3 is what percent of 4?
3 is ?% of 4

We don't know both sides of the *of* so we divide the number closest to the *of* into the other number.

$$\begin{array}{r} 0.75 \\ 4\overline{)3.00} \end{array} \quad 0.75 = 75\%$$

7 is 75% more than 4.

Let's see if that makes sense. The percent problem that you have been used to solving is: What is 75% more than 4?
  The hard way: 75% of 4 is $0.75 \times 4 = 3$
    4 plus an increase of 3 is 7.
  The easy way: 75% more means the original 100% plus 75% more. $100\% + 75\% = 175\%$.
    $4 \times 1.75 = 7$

728. What is 15% more than 60?

  The hard way: 15% of 60 is $0.15 \times 60 = 9$ The gain is 9.
    Add the gain to the original 60: $60 + 9 = 69$

  The easy way: A 15% gain means the original 100% plus 15% more. $100\% + 15\% = 115\%$.
    $60 \times 1.15 = 69$

729–731 Complete Solutions and Answers

729. Jan had charged an equal amount on her seven credit cards (Disaster Card, Pleasa Card, American Distress, Uncover Card, etc.) and she had borrowed $100 from you. The total came to $21,800. How much did she owe on each credit card?

Let x = the amount she owed on each credit card.
Then 7x = the amount she owed on all seven cards.
Then 7x + 100 = the total amount she owed.

The equation then is   7x + 100 = 21,800
  Subtract 100 from both sides   7x = 21,700
  Divide both sides by 7   x = 3,100

Jan owed $3,100 on each card.

730. Find the LCD of $\frac{1}{4}$ and $\frac{1}{8}$ and $\frac{1}{24}$

The smallest number that 4, 8, and 24 all divide evenly into is 24.

731. Ivy wasn't sure what to do with that cylinder (50 feet tall and with a diameter of 20 feet). She decided to fill it with "leftover" pennies that she received when selling ice cream. There are approximately 50,000 pennies in a cubic foot. Roughly, how many pennies will that cylinder hold?

$$V_{cylinder} = \pi r^2 h \text{ becomes } V = \pi(10^2)(50) \doteq 15{,}708 \text{ ft}^3$$

(We can round because we are given that there are *approximately* 50,000 pennies in a cubic foot. Our final answer will only be approximate.)

We now have 15,708 cubic feet, which we want to convert into pennies. There are two possible conversion factors:

$$\frac{50{,}000 \text{ pennies}}{1 \text{ ft}^3} \quad \text{or} \quad \frac{1 \text{ ft}^3}{50{,}000 \text{ pennies}}$$

We chose the conversion factors that makes the units cancel.

$$\frac{15708 \text{ ft}^3}{1} \times \frac{50{,}000 \text{ pennies}}{1 \text{ ft}^3}$$

$$= 785{,}400{,}000 \text{ pennies.}$$

I don't convert this to $7,845,000 because the question asked "how many pennies?"

Complete Solutions and Answers 732–735

732. Ivy put 18 (identical) cobs of *Zea mays* into a 60 gram bucket. The whole thing weighed 834 grams. How much did each cob weigh?

Let x = the weight of one cob.
Then 18x = the weight of all the corn.
Then 18x + 60 = the weight of the corn plus the bucket.

The equation becomes $\quad\quad$ 18x + 60 = 834
Subtract 60 from both sides $\quad\quad$ 18x = 774
Divide both sides by 18 $\quad\quad\quad\quad$ x = 43

Each cob weighed 43 grams.

733. If set A = {red, yellow, green} and set B = {orange, blue}, then does card A + card B = card (A ∪ B)?

card A = 3 $\quad$ since there are three members of {red, yellow, green}
card B = 2
card (A ∪ B) = card {red, yellow, green, orange, blue} = 5

Yes. card A + card B = card (A ∪ B) is true.
$\quad\quad\quad$ 3 $\quad$ + $\quad$ 2 $\quad$ = $\quad\quad$ 5

734. You drive $\frac{3}{8}$ of a mile to the New Mall. You park and walk $1\frac{3}{4}$ miles to the nearest store. How far have you traveled?

$\frac{3}{8} + 1\frac{3}{4}$

$= \frac{3}{8} + 1\frac{6}{8}$

$= 1\frac{9}{8} = 2\frac{1}{8}$ miles

735. Jan has pink hair 5.37% of the time. What percent of the time does she not have pink hair?

100% − 5.37%
= 1 − 0.0537
= 0.9463
= 94.63% of the time she doesn't have pink hair.

> Or even shorter... $\quad$ 100% − 5.37% = 94.63%

| 754–756 | **Complete Solutions and Answers**

754. A 12-hour shift. $26 per hour. The government would take out 20% for income taxes, 6% for medical insurance, 5% for Social Security, and 7% for unemployment insurance. How much would Jan get per day?

12 hours at $26 per hour = $312 gross pay.

20% + 6% + 5% + 7% = 38%
62% would be left for Jan. (100% − 38% = 62%)
62% of $312
0.62 × 312 = $193.44 net pay per day

Jan was delighted. She could pay her rent. She could buy makeup and some new clothes.

755. Jan had had two cats. She now had one. What percent change had she experienced?

She lost one cat.
One cat is what percent of two cats?
1 = ?% of 2
$1 \div 2 = \frac{1}{2} = 50\%$

When Jan went from one cat to two cats, she gained 100%. When she went from two cats to one cat, she lost 50%.

A 100% gain matches up with a 50% loss.

756. If you start with 5 quarts of blood, how much would you have to lose to have lost 40%? Do this problem in your head and just write down the answer. No calculating with a pencil. No need for a calculator.

How do you do 40% of 5 in your head? You know that it's going to be 40% times 5 (since you know both sides of the *of*.)

Without paying attention to the decimals, you know that it will be something like 40 times 5. That's 200. The answer could be any of these:

0.2 or 2 or 20 or 200

It couldn't be 20 or 200, because you were only starting out with 5 quarts of blood.
It is either 2 or 0.2.
40% of 5 is roughly half of 5. (50% of 5 would be exactly half.)
2 quarts lost out of 5 makes sense.

Complete Solutions and Answers | 757–760

757. Jan started her work as assistant manager at Ivy's Ice Cream. A full container (5 gallons) of Cloudy Vanilla weighs 36 pounds. Two and a half quarts had already been scooped out. How many gallons and quarts were left?

$$\begin{array}{r} 5 \text{ gallons} \\ - 2\tfrac{1}{2} \text{ quarts} \\ \hline \end{array} \qquad \begin{array}{r} 4 \text{ gallons } 4 \text{ quarts} \\ - 2\tfrac{1}{2} \text{ quarts} \\ \hline 4 \text{ gallons } 1\tfrac{1}{2} \text{ quarts} \end{array}$$

758. The cardinal numbers are the numbers used to count the members of a set. (**card A** means the cardinal number associated with set A.)

#1 = {chocolate, **SHIVERING STRAWBERRY**}
#2 = {Vacant Vanilla, Crunchy Cinnamon}
#3 = {**SHIVERING STRAWBERRY**, Crunchy Cinnamon}
#4 = {Vacant Vanilla, chocolate, Crunchy Cinnamon}

Is it true that: card #1 = card #2 = card #3 = card #4?

card #1 = 2
card #2 = 2
card #3 = 2
card #4 = 3

It is not true that card #1 = card #2 = card #3 = card #4.
(It is true that card #1 = card #2 = card #3.)

759. As they got out of the car, nature helped the healing process. Joe threw up. He had weighed 174 pounds, 4 ounces. He now weighed 170 pounds, 7 ounces. How much did he lose? (1 pound = 16 ounces)

$$\begin{array}{r} 174 \text{ lbs. } 4 \text{ oz.} \\ - 170 \text{ lbs. } 7 \text{ oz.} \\ \hline \end{array} \qquad \begin{array}{r} 173 \text{ lbs. } 16 \text{ oz.} + 4 \text{ oz.} \\ - 170 \text{ lbs. } 7 \text{ oz.} \\ \hline \end{array} \qquad \begin{array}{r} 173 \text{ lbs. } 20 \text{ oz.} \\ - 170 \text{ lbs. } 7 \text{ oz.} \\ \hline 3 \text{ lbs. } 13 \text{ oz.} \end{array}$$

Joe lost 3 pounds, 13 ounces and felt much better.

760. Name the four subsets of {Cassie, Charlie}.

{ }, {Cassie}, {Charlie}, and {Cassie, Charlie}.

773–774 | Complete Solutions and Answers

773. It was 3.2 miles to the nearest hospital. You drove Jan at 10.2 mph (miles per hour). How long would it take to get there? Give your answer to the nearest minute.

Let t = the number of hours to get to the hospital. (We are using *hours* because the given information is in miles and miles per hour.)

$$d = rt \text{ becomes} \qquad 3.2 = 10.2t$$
Divide both sides by 10.2 $\qquad 0.3137 \text{ hours} \approx t$

To convert hours into minutes we use a conversion factor.

$$\frac{0.3137 \text{ hours}}{1} \times \frac{60 \text{ minutes}}{1 \text{ hour}} \approx 18.82 \doteq 19 \text{ minutes}$$

The other way to do this problem is to first convert 10.2 mph into miles per minute. Then we could let t = the number of minutes to get to the hospital. Either way, we would get the same answer.

I don't believe you.
Why not?
Because I'm feeling ornery.*
Okay.

Let t = the number of minutes to get to the hospital.

$$\frac{10.2 \text{ miles}}{1 \text{ hour}} \times \frac{1 \text{ hour}}{60 \text{ minutes}} = 0.17 \text{ miles per minute}$$

$$d = rt \text{ becomes} \qquad 3.2 = 0.17t$$
Divide both sides by 0.17 $\qquad 18.82 \approx t$

t ≐ 19 minutes.
I feel better now. Thank you.
You are welcome.

774. Jan scooped out a ball (a sphere) of *Matrimonial Maple* and gave it to her. It had a radius of three-fifths of an inch. What was the volume of that ball of ice cream? $V_{sphere} = \frac{4}{3}\pi r^3$ (Use 3 for π in this problem.)

$$V = \frac{4}{3}\pi r^3 \approx \frac{4}{3} \times 3 \times \left(\frac{3}{5}\right)^3 = \frac{4}{3} \times 3 \times \frac{3}{5} \times \frac{3}{5} \times \frac{3}{5} = \frac{108}{125} \text{ cubic inches}$$

---

* Ornery is what my mother used to call crabby. It's what my grandfather called crotchety.

Complete Solutions and Answers 782–783

782. $(7.8)^2 = 60.84$

```
      7.8
   ×  7.8
     ────
      624
      546
     ────
     6084    ⟹  60.84
```

783. The natural numbers = {1, 2, 3, 4, 5, . . .}. If y is a natural number, how many solutions would $3y + 7 < 20$ have?

Let's count them.
Does y = 1 work? $3y + 7 < 20$ becomes $3 + 7 < 20$. True.
Does y = 2 work? $3y + 7 < 20$ becomes $6 + 7 < 20$. True.
Does y = 3 work? $3y + 7 < 20$ becomes $9 + 7 < 20$. True.
Does y = 4 work? $3y + 7 < 20$ becomes $12 + 7 < 20$. True.
Does y = 5 work? $3y + 7 < 20$ becomes $15 + 7 < 20$. False.
Does y = 6 work? $3y + 7 < 20$ becomes $18 + 7 < 20$. False.
Does y = 7 work? $3y + 7 < 20$ becomes $21 + 7 < 20$. False.

The numbers 1, 2, 3, and 4 make $3y + 7 < 20$ true. $3y + 7 < 20$ has four solutions.

If we didn't say that y had to be a natural number, there would be a zillion solutions.
If y = 1.4, then $3y + 7 < 20$ becomes $4.2 + 7 < 20$. True.
If y = 1.472, then $3y + 7 < 20$ becomes $4.416 + 7 < 20$. True.

Later, when we get to algebra, we are going to solve inequalities such as $3y + 7 < 20$ in a similar way* to solving equalities such as $3y + 7 = 20$.

For example,	$3y + 7 < 20$	
Subtract 7 from both sides	$3y < 13$	
Divide both sides by 3	$y < 13/3$	$y < 4\frac{1}{3}$

---

* In a similar way, but not exactly the same. For my super advanced readers, we note that in solving inequalities in algebra, whenever we multiply or divide by a negative number, we must switch the sense of the inequality: < will become >. For now, please don't pay attention to this footnote.

| 789–794 | Complete Solutions and Answers |

789. Yesterday, I was reading this.

> In short, the chart below puts the lie to the alleged virtuous circle of Keynesian stimulus. There has been no pump-priming of consumption spending, production, jobs, income, and more of the same.
>
> Indeed, the Fed's balance sheet has grown by *900%* during the last 16 years—from $500 billion to $4.5 trillion. By contrast, labor hours have risen by only *6.7% on a cumulative basis*.

Since I had bought the book, I reached for a pencil and scratched out one of the numbers. Which one?

To go from $500 billion to $4.5 trillion means going from a half trillion dollars to 4.5 trillion dollars.

That's an increase of 4 trillion dollars.

4 trillion is what percent of one-half trillion dollars?

4 = ?% of ½

$$4 \div \tfrac{1}{2} = \tfrac{4}{1} \times \tfrac{2}{1} = 8 = 800\%$$

An increase of 100% means doubling the original amount.
An increase of 200% means tripling the original amount.
An increase of 800% means 9x the original amount.
I'm not going to mention the name of the author because I don't want to embarrass him/her. The author's initials are D.S.

790. Suppose there was a 0.043% chance that *they* would knock on Jan's door. Round that to the nearest percent.

0.043% rounded to the nearest percent is 0%.
If it were 0.53%, that would round to 1%.
If it were 99.4%, that would round to 99%.
If it were 99.502%, that would round to 100%. 100% is certainty.

794. The little girl bought it for $46 and wanted to make a 20% profit. How much was the little girl willing to sell her camera for?

20% more than $46 can be done in two ways:

The harder way:
        20% of $46 is 0.2 × 46 = $9.20
        Adding $9.20 to $46 yields $55.20

The easier way:
        20% plus the original 100% is 120%
        120% of $46 is 1.2 × 46 = $55.20

Complete Solutions and Answers | 800–802

800. In the previous chapter we learned that a bucket (cylinder) of Ralph's eyelashes was about 50 cubic inches. It cost $41.

In the previous problem we learned that a cone-of-lashes had a volume of 37 cubic inches. Using a conversion factor, find out how much Ralph should charge for his cone-of-lashes.

Since 50 cubic inches is $41, the conversion factor will be either

$$\frac{50 \text{ cubic inches}}{\$41} \quad \text{or} \quad \frac{\$41}{50 \text{ cubic inches}}$$

$$\frac{37 \text{ cubic inches}}{1} \times \frac{\$41}{50 \text{ cubic inches}}$$

$$\frac{37 \text{ cubic inches}}{1} \times \frac{\$41}{50 \text{ cubic inches}}$$

= $30.34 is what Ralph should charge for a cone-of-lashes.

801. When Darlene works, she makes $8 an hour. How long would it take her to earn $500?

Since $8 = 1 hour, the conversion factor will be either $\frac{\$8}{1 \text{ hr}}$ or $\frac{1 \text{ hr}}{\$8}$

$$\frac{\$500}{1} \times \frac{1 \text{ hr}}{\$8} = \frac{500 \text{ hr}}{8} = 62.5 \text{ hours} = 62 \text{ hours and 30 minutes}$$

802. Change $\frac{21}{24}$ into a percent.

To make things easier, we first reduce the fraction to $\frac{7}{8}$

If you remember the nine conversions that you were asked to memorize in *Life of Fred: Decimals and Percents*, you simply write 87½%.

Otherwise, you have to divide it out:

```
    0.875
8) 7.000
    64
    ‾‾
    60
    56
    ‾‾
    40
    40
```

0.875 = 87.5% or 87½%

219

| 803 | Complete Solutions and Answers |

803. (continuing the previous two problems) How much would a ton of sparklettes cost?

We know: 1 ton = 2,000 pounds

Each sparklette cost 17¢.

Each sparklette weighs $\frac{3}{8}$ pounds.

We want to convert one ton into cost. We have three conversion factors to use.

$$\frac{1 \text{ ton}}{1} \times \frac{2000 \text{ lbs.}}{1 \text{ ton}} \times \frac{\text{each sparklette}}{3/8 \text{ lb.}} \times \frac{17¢}{\text{each sparklette}}$$

Using the fact that 3/8 = 0.375, which is one of the nine conversion factors,

$$= 2000 \div 0.375 \times 17 \approx 90{,}666.66 \doteq 90{,}666¢ = \$906.66$$

Slow discussion:
Question: We started with a ton. How did we know which conversion factor to use?
Answer: Only one of the three known facts involved tons. It was that one ton = 2,000 pounds.

Second question: Once we had converted one ton into pounds using the $\frac{2000 \text{ lbs.}}{1 \text{ ton}}$ conversion factor, how did we know which conversion factor to use next?
Second answer: After we used that conversion factor, we had pounds. Looking at the three known facts, two of them involved pounds. The "one ton = 2000 pounds" would have taken us back into tons. That's the wrong direction. We used the other fact: Each sparklette weighed three-eighths of a pound. That's how we got the second conversion factor.

We could have broken this problem up into three smaller problems.

First, $\frac{1 \text{ ton}}{1} \times \frac{2000 \text{ lbs.}}{1 \text{ ton}}$ to get tons converted into pounds.

Second, $\times \frac{\text{each sparklette}}{3/8 \text{ lb.}}$ to convert pounds into number of sparklettes.

Third, $\times \frac{17¢}{\text{each sparklette}}$ to convert number of sparklettes into cents.

Complete Solutions and Answers | 804–806

804. Let x = the cost (in cents) of one scoop of ice cream. The cost of a cone is 18¢. The kids had 9 scoops and 4 cones. The total bill was $4.50. How much is the cost of one scoop of ice cream?

The first step of the word problem has already been done: Let x = the cost (in cents) of one scoop of ice cream.

Then 9x = the cost of 9 scoops.
72¢ = the cost of 4 cones.   (4 × 18 = 72)
Then 9x + 72 = the total bill.
450¢ = the total bill.   We are going to work in cents.

The equation	9x + 72 = 450
Subtract 72 from both sides	9x = 378
Divide both sides by 9	x = 42

Each scoop of ice cream costs 42¢.

805. Jan had started her shopping spree with $193.44 in her pocket. She spent $4.25 on sparklettes. She spent $150.83 on cat food. She spent the rest on happy gingerbread men.

What percent of her original $193.44 did she spend on these happy men?

Give your answer to the nearest percent.

She spent $155.08 on sparklettes and cat food. ($4.25 + $150.83)
She spent $38.36 on gingerbread men. ($193.44 − $155.08)

$38.36 is what percent of $193.44?
38.36 = ?% of 193.44
38.36 ÷ 193.44 ≈ 0.1983 = 19.83% ≐ 20%

Jan spent about twenty percent of her paycheck on these happy men.

806. For every million knocks on Jan's door, 0.43% of them would be Hollywood producers. How many is that?

What is 0.043% of a million?
0.043% of 1,000,000 = ?
You know both sides of the *of* so you multiply.
0.043% × 1,000,000
0.00043 × 1,000,000
= 430 knocks on her door would be movie producers.

> 824–828

Complete Solutions and Answers

824. *How many* subsets does {Cassie, Charlie, Jan, Ivy} have?

    We know that {Cassie, Charlie, Jan} has 8 subsets. Take each of those 8 subsets and add in Ivy. For example, {Charlie, Jan} would become {Charlie, Jan, Ivy}. That makes 8 more subsets.

    {Cassie, Charlie, Jan, Ivy} has 16 subsets. (Hint: 8 + 8 = 16)

825. Jan traded 18 boxes of macaroni and cheese for a gas mask. If 3 boxes of m & c weighed 19 ounces, how much did all 18 boxes weigh?

$$\frac{18 \text{ boxes}}{1} \times \frac{19 \text{ ounces}}{3 \text{ boxes}}$$

$$= \frac{\overset{6}{\cancel{18 \text{ boxes}}}}{1} \times \frac{19 \text{ ounces}}{\underset{1}{\cancel{3 \text{ boxes}}}}$$

$$= 114 \text{ ounces of macaroni and cheese}$$

826.

SPEAKING PARTS

Lysander	Alex
Demetrius	Bob
Hermia	Cassie
Helena	Diana
Oberon, King of Fairies	Edgar
Titania, Queen of Fairies	Fay
Puck, Oberon's jester	Mickey R.
Front end of donkey	Gordo

NON-SPEAKING PARTS

Donkey butt	Jan

These two sets—the speaking parts and the non-speaking parts—are all the people acting in the play. The union of these two sets is equal to all the actors in the play. Are the two sets disjoint?

    Since no actor is in both sets, the sets are disjoint.

827. Solve $6x + 43 = 145$

    Subtract 43 from both sides $6x = 102$
    Divide both sides by 6 $x = 17$

828. How many subsets does {A, B, C, D, E, F} have?

    We know that {A} has 2 subsets.   { } and {A}
    We know that {A, B} has 4 subsets. We did that two problems ago.
    We know that {A, B, C} has 8 subsets. We did that in the previous problem.
    {A, B, C, D} has 16 subsets.
    {A, B, C, D, E} has 32 subsets.
    {A, B, C, D, E, F} has 64 subsets.

Complete Solutions and Answers | 829–830

829. As everyone in business knows, your profit equals your income minus your expenses. P = I – E
Does a business owner want: I > E, I = E, or I < E?

If I > E, then your income is greater than your expenses, and you are making a profit. You can stay in business.

If I = E, then your income matches your expenses, and you will have no money to live on. You can't pay your rent. You can't buy groceries. You can't even ride the bus.

If I < E, you are in the situation that Ivy was in before Jan started singing and dancing. Ivy's savings were being eaten up by the mortgage on the store, government taxes, the cost of making the thousand flavors of ice cream, and so on. After Ivy used up all her savings, she would have to close the store.

830. In order to get into Ivy's and see Jan perform, you have to buy 6 cartons of aNchoVY-LaMB With a RiBBoN oF YaM and a $5 plastic spoon. The whole package costs $24.62.

How much does a carton of aNchoVY-LaMB With a RiBBoN oF YaM cost?

You want to learn what a carton of ice cream costs, so you . . .
Let x = the cost of a carton of aNchoVY-LaMB With a RiBBoN oF YaM ice cream.
Then 6x = the cost of six cartons
Then 6x + 5 is the cost of six cartons plus a plastic spoon.
We know that the whole package costs $24.62.

The equation is $6x + 5 = 24.62$
If you haven't read <u>Zillions of Practice Problems Pre-Algebra 0 with Physics</u> solving this equation may be new to you.

The whole point is that we do the same thing to both sides of the equation.

The equation is	$6x + 5 = 24.62$
Subtract 5 from both sides	$6x = 19.62$
Divide both sides by 6	$x = 3.27$

A carton of aNchoVY-LaMB With a RiBBoN oF YaM ice cream costs $3.27

| 835–838 | Complete Solutions and Answers |

835. Ralph's cone-of-lashes normally would cost $30.34. He decided to sell it with a 19% discount. To the nearest cent, how much would he charge?

If there is a 19% discount, then the new price is 81% of the old price. (100% − 19% = 81%)
new price = 81% of old price
new price = 0.81 of 30.34
We know both sides of the *of*, so we multiply.
new price = 24.5754 ≐ $24.58 is the new discounted price.

836. Beth shook her head and said, "The set of all patients we have here at this veterinary hospital and the set of all human beings are __disjoint__."

A veterinary hospital treats only animals. Animal doctors are often called vets. Vet is short for veterinarian. Most people can spell vet, but can't spell veterinarian.

837. If sets C and D are disjoint, will it always be true that
card C + card D = card (C ∪ D)?

In order to see whether this is true, you have to mentally try out various possibilities.
There are many different approaches to this problem. The way that I think about it is to think about putting the members of set C in my left hand and the members of set D in my right hand. (That makes the sets disjoint.)
I count the members of C in my left hand. I count the members of D in my right hand. Suppose there were 100 in my left hand and 57 in my right hand.
If I combine the sets by putting my hands together, I'm going to have 157. I won't lose any or gain any. No surprises.
It is true that if C and D are disjoint, then card C + card D will always equal card (C ∪ D).

838. Jan thought of the perfect new ice cream flavor: Bison-Blueberry. She phoned Ivy to tell her. She called at 1:20 a.m. Ivy had gone to sleep at 9:50 p.m. How long had Ivy been sleeping before Jan woke her up?

9:50 p.m. to 10 p.m. (→ 10 minutes)   10 p.m. to 1 a.m. (→ 3 hours)
1 a.m. to 1:20 a.m. (→ 20 minutes)
3 hours and 30 minutes

Complete Solutions and Answers | 855–857

855. Solve $8x + 5 = 31$

We start with	$8x + 5 = 31$
Subtract 5 from both sides	$8x = 26$
Divide both sides by 8	$x = 3.25$

856. How much larger is $8\frac{1}{4}$ than $3\frac{7}{8}$?

Using the **General Rule** (If you don't know whether to add, subtract, multiply, or divide, first restate the problem with really simple numbers), we might ask how much larger is 5 than 2? The answer is 3. We subtracted.

How much larger is $8\frac{1}{4}$ than $3\frac{7}{8}$? means $8\frac{1}{4} - 3\frac{7}{8}$

$$8\frac{1}{4} \qquad 8\frac{2}{8} \qquad 7\frac{8}{8} + \frac{2}{8} \qquad 7\frac{10}{8}$$
$$-3\frac{7}{8} \qquad -3\frac{7}{8} \qquad -3\frac{7}{8} \qquad -3\frac{7}{8}$$
$$ 4\frac{3}{8}$$

In an hour Ivy could fill $4\frac{3}{8}$ more feet of order forms than Darlene. Ivy fired Darlene. She wasn't worth $8 an hour.

857. Darlene ate three-eighths of her $\frac{108}{125}$ cubic inch ball of ice cream and said that she was full. She gave the rest of it to Joe who eagerly devoured it. How many cubic inches of *Matrimonial Maple* did Joe eat?

If she ate $\frac{3}{8}$ of it, she left $\frac{5}{8}$ of it for Joe.

$\frac{5}{8}$ of $\frac{108}{125}$

$$\frac{5}{8} \times \frac{108}{125} = \frac{\overset{1}{\cancel{5}}}{\underset{4}{\cancel{8}}} \times \frac{\overset{54}{\cancel{108}}}{\underset{25}{\cancel{125}}} = \frac{54}{100} = \frac{27}{50}$$

Joe ate $\frac{27}{50}$ of a cubic inch of *Matrimonial Maple*.

Complete Solutions and Answers

858. You can run 8 mph (miles per hour). Jan can run 26.8% faster than you. How fast can Jan run? (Round your answer to the nearest tenth of a mile per hour.)

This is a "15% more than" problem. There are two ways to do it.
One way: 26.8% of 8 is $0.268 \times 8 = 2.144$
Add 2.144 to the original 8 and get $10.144 \doteq 10.1$ mph

Another way: 26.8% more than means 126.8% (26.8% + 100%) of the original amount. 126.8% of 8 = $1.268 \times 8 = 10.144 \doteq 10.1$ mph

859. Ivy is no dummy. She invests most of her profits rather than spending them. If she invests a million dollars and it earns 5% per year, how much will she have in a year?

We want to compute 5% more than $1,000,000.
This is a "x% more than" problem. There are two ways to do it.
One way: 5% of 1,000,000 is 50,000
Add 50,000 to the original 1,000,000 and get $1,050,000.

Easier way: 5% more than means 105% (100% + 5%) of the original amount. 105% of 1,000,000 = $1.05 \times 1,000,000 = \$1,050,000$.

The Easier Way really is much easier when you have to compute how that million dollars will grow over several years.

Original amount	1,000,000	
After one year	$1,000,000 \times 1.05 =$	$1,050,000
After two years	$1,000,000 \times 1.05 \times 1.05 =$	$1,102,500
After three years	$1,000,000 \times (1.05)^3 =$	$1,157,625
After four years	$1,000,000 \times (1.05)^4 \doteq$	$1,215,506

($\doteq$ means "equals after rounding)

860. It wasn't Jan.

861. Which is larger 3^2 or 2^3?

Three squared $= 3^2 = 3 \times 3 = 9$.
Two cubed $= 2^3 = 2 \times 2 \times 2 = 8$.
$3^2 > 2^3$

Complete Solutions and Answers | 862–866

862. $(4\frac{1}{6})^2 = 4\frac{1}{6} \times 4\frac{1}{6} = \frac{25}{6} \times \frac{25}{6} = \frac{625}{36} = 17\frac{13}{36}$

863. For what values of x will 1^x equal one?

When x = 23, 1^x becomes 1^{23}, which equals 1 × 1 × 1 × 1 × 1 × 1 × 1 × 1 × 1 × 1 × 1 × 1 × 1 × 1 × 1 × 1 × 1 × 1 × 1 × 1 × 1 × 1 × 1, which equals one.

When x = 51, 1^x becomes 1^{51}, which equals 1 × 1 × 1 × 1 × 1 × 1 × 1 × 1 × 1 × 1 × 1 × 1 × 1 × 1 × 1 × 1 × 1 × 1 × 1 × 1 × 1 × 1 × 1 × 1 × 1 × 1 × 1 × 1 × 1 × 1 × 1 × 1 × 1 × 1 × 1 × 1 × 1 × 1 × 1 × 1 × 1 × 1 × 1 × 1 × 1 × 1 × 1 × 1 × 1 × 1 × 1 which equals one.

For every value of x, 1^x will equal one.

In algebra you will learn that 1^0 equals one, that 1^{-3} equals one, that $1^\pi = 1$. We really mean it when we say that for every value of x, $1^x = 1$.

864. If you put a fake eyelash on your pinna (your ear), wouldn't it be called an earlash?

865. One of her cats, Juno, was chewing on your ankle. The other 20 cats had leaped out the window. What percent of Jan's cats remained in her apartment?

One is what percent of 21?
1 = ?% of 21
$1 \div 21 = \frac{1}{21} = 0.04761904761904761904761 \doteq 4.8\%$

866. You had $2,000 in your checking account. The final bill was for 80% of the $2,000. How much was the bill?

80% of 2,000. It is multiplication. Ignoring the decimals, the answer will be one of these:
  16
  160
  1,600
  16,000
  160,000

Eighty percent of your checking account will be most of that account. $1,600 is the only one that makes sense. $1,600 is most of $2,000.

| 870–873 | Complete Solutions and Answers

870. Solve $8y + 31 = 5y + 82$

We start with	$8y + 31 = 5y + 82$
Subtract 31 from both sides	$8y = 5y + 51$
Subtract 5y from both sides	$3y = 51$
Divide both sides by 3	$y = 17$

871. If Jan received 4 knocks on her door every week, how long would it take to receive a million knocks?

Four knocks match up with one week. The conversion factor will be either $\dfrac{4 \text{ knocks}}{\text{one week}}$ or $\dfrac{\text{one week}}{4 \text{ knocks}}$

$$\dfrac{1{,}000{,}000 \text{ knocks}}{1} \times \dfrac{\text{one week}}{4 \text{ knocks}}$$

$$= \dfrac{1{,}000{,}000 \text{ knocks}}{1} \times \dfrac{\text{one week}}{4 \text{ knocks}}$$

$$= 250{,}000 \text{ weeks}$$

For fun, we'll convert that into years:

$$\dfrac{250{,}000 \text{ weeks}}{1} \times \dfrac{1 \text{ year}}{52 \text{ weeks}} \doteq 4{,}800 \text{ years}$$

872. Solve $\quad 4x + 6x + 18 = 27$

Combine like terms	$10x + 18 = 27$
Subtract 18 from both sides	$10x = 9$
Divide both sides by 10	$x = \dfrac{9}{10}$ or 0.9

873. How many minutes did they walk? Use a conversion factor.

We want to convert 660 seconds into minutes.

$$\dfrac{660 \text{ seconds}}{1} \times \dfrac{1 \text{ minute}}{60 \text{ seconds}}$$

$$= \dfrac{660 \text{ seconds}}{1} \times \dfrac{1 \text{ minute}}{60 \text{ seconds}}$$

$$= 11 \text{ minutes they walked}$$

Complete Solutions and Answers | 880–891

880. It was 9 a.m. Jan's shift (and job) were over. She bought a Bison-Blueberry pie with a scoop of Vacant Vanilla on top. $2.42 (= $2 + 42¢)

She sat at one of the outdoor tables at Ivy's and ate it. That was a meal for Jan. If she kept spending $2.42, and she started with $93.44, how many meals would she have?

How many $2.42 meals can you get if you start with $93.44? Using the **General Rule** (If you don't know whether to add, subtract, multiply, or divide, first restate the problem with really simple numbers), we might ask how many $2 meals could you buy with $8? You could buy 4 meals. We divided.

$93.44 \div 2.42 \approx 38.61$

Since you can't buy part of a meal, the answer is 38 meals.

888. A gallon is 231 cubic inches. A 5-gallon container (a cylinder) of Cloudy Vanilla has a radius of 6 inches. To the nearest inch, how tall is that container? (Use 3 for π.)

The volume is 5 × 231 = 1,155 cubic inches for a 5-gallon container.

$V_{cylinder} = \pi r^2 h$ $1155 = (3)(6^2)h$
                                  $1155 = 108h$

Divide both sides by 108 $10.69444 \approx h$
Round $11 \doteq h$

The height of the can of Cloudy Vanilla is about 11 inches.

891. If the little girl sold her camera for $62, what would be her percent profit? (Round to the nearest percent.)

She would have made a profit of $16. ($62 − $46)
$16 is what percent of $46?
16 = ?% of 46
$16 \div 46 = \dfrac{16}{46} \approx 0.347826 = 34.7826\% \doteq 35\%$

| 900–902 | **Complete Solutions and Answers** |

900. One of the most famous formulas from algebra is d = rt (distance equals rate times time). Ivy's store is 52 feet wide.
Ivy bought a special robot to carry Puffy Buffy ice cream, because it was so dangerous. It could run from the left side of the store to the right side at 8 feet per second. How long would that take?

Let t = the number of seconds that it would take for the robot to run the width of the store.

    d = rt becomes   52 = 8t
    Divide both sides by 8   6.5 = t

It would take 6.5 seconds.

In the old days before calculators you would have to write:

$$\begin{array}{r} 6.5 \\ 8{\overline{\smash{)}52.0}} \\ \underline{48} \\ 40 \\ \underline{40} \end{array}$$

and that might have taken less time that it would take to haul out a calculator, turn it on, and punch in the numbers.

901. Let P be the set of all pizzas in the world right now. Find two sets, A and B, such that A ∪ B = P where A and B are disjoint.

    There are many possible answers. Your answer will probably be different than mine.
My first answer: Let A be the set of all pizzas that contain anchovies. Let B be the set of all pizzas that don't contain anchovies.
My second answer: Let A be the set of all pizzas made by Stanthony. Let B be all the other pizzas.
My third answer: Let A be all the pizzas that are now in Kansas. Let B be all the pizzas that are not now in Kansas.

902. Let set A = the set of all possible things to do on an afternoon. Let set E = the set of activities involving eating. Let B = the activity of bowling. Does A = E ∪ B?

    ∪ stands for union. Is eating or going bowling the only two things that you and Jan could be doing this afternoon? Of course not. A ≠ E ∪ B
    ≠ means "not equal."

Complete Solutions and Answers

917. The director told Jan she would be paid $44 after tomorrow's performance. Jan needed the money right now. Edgar said he would lend her $35.20 right now and she could pay him $44 tomorrow. The interest charge would be $8.80. What percent interest is that?

$8.80 is what percent of $35.20 that Jan borrowed?
8.80 = ?% of 35.20

You don't know both sides of the *of* so you divide the number closest to the *of* into the other number.

$$8.80 \div 35.20 = \frac{8.80}{35.20} = 0.25 = 25\%$$

Of course, this is 25% for money that is borrowed for one day. If we multiply that times the 365 days* in the year, we find that the annual interest rate would be 9125% per year.

918. He was 162 pounds. He is now 18 pounds, which is the weight of his ashes. What fraction of his original weight is he now?

18 pounds is what fraction of 162?
18 = ? of 162

We can use the same method as the one you have used for years with percents. If you don't know both sides of the *of*, you divide the number closest to the *of* into the other number.

$162 \overline{\smash{)}18}$ which is $\frac{18}{162}$ which reduces to $\frac{1}{9}$

If we had tried to do this problem using percents, we would have had an answer of 11.1111111111111111111111111111111111111111111111111%, which isn't as pretty as 1/9.

---

* I know that there are roughly 365¼ days in a year, but I just wanted to get a rough idea what the annual rate is. There are all kinds of years: seasonal years, the fiscal years, the academic years, etc. For the astronomical year, we have to divide the number of spins of the earth into the time it takes for the earth to go around the sun. It doesn't come out evenly. One rough estimate is 365.242199 days in a year.

| 919–922 | Complete Solutions and Answers |

919. A cylinder of Ralph's eyelashes is 4 inches tall with the radius of 2 inches. What is its volume? $V_{cylinder} = \pi r^2 h$ Use 3.1 for π and round your answer to the nearest cubic inch.

$$V = \pi r^2 h \approx (3.1)2^2(4) = 49.6 \doteq 50 \text{ cubic inches}$$

($\approx$ means "approximately equal to" and $\doteq$ means "equals after rounding")

920. Is 1003rd a cardinal number?

Cardinal numbers are used to count the number of members in a set. The number of members in that set of 1,003 ice cream flavors is the cardinal number 1,003.

1003rd is an ordinal number, not a cardinal number.

There could be 1,003 runners in a race (cardinal number), but you wouldn't want to be the one who came in 1003rd in that race (ordinal number).

921. Three-eights is what percent of $\frac{15}{16}$?

$$\frac{3}{8} = ?\% \text{ of } \frac{15}{16}$$

We don't know both sides of the *of,* so we divide the number closest to the *of* into the other number.

$$\frac{3}{8} \div \frac{15}{16} = \frac{3}{8} \times \frac{16}{15} = \frac{3}{8} \times \frac{16}{15} = \frac{2}{5} = 0.4 = 40\%$$

922. You and Jan head off to Stanthony's PieOne pizza place for lunch. You have a pepperoni pizza, a salad, a scoop of anchovy-lamb with a ribbon of yam on an ice cream cone and half dozen other items. Name the one thing that you had that wasn't once alive.

It wasn't the pepperoni, which is meat. It wasn't the pizza crust, which is made of wheat. It wasn't the lettuce in the salad. It wasn't the eggs in the salad. (It was a chef's salad, which normally has hard boiled eggs in it.) It wasn't the vinegar on the salad. It wasn't the olive oil in the salad dressing. It wasn't the anchovy (a fish). It wasn't the lamb. It wasn't the yam. It wasn't the ice cream cone (made from wheat). It was the glass of water.

Complete Solutions and Answers | 930–939

930. If sets E and F are *not* disjoint, which of these will be true:

    i) card E + card F $<$ card (E ∪ F) $<$ means "less than"
    ii) card E + card F $=$ card (E ∪ F)
    iii) card E + card F $>$ card (E ∪ F) $>$ means "greater than"

Suppose set E = {☎, ☺, ✈}   Then card E = 3
Suppose set F = {☎, ♣}   Then card F = 2

    Note that E and F are not disjoint since they both have ☎ in them.
    card E + card F = 5   card (E ∪ F) = card {☎, ☺, ✈, ♣} = 4
    iii) card E + card F $>$ card (E ∪ F) is true.

932. How long might Jan expect to wait to receive a knock on the door from a movie producer?

From the previous two problems, we know that 250,000 weeks matches up with a million knocks, which matches up with 430 movie producers.

    250,000 weeks = 1,000,000 knocks = 430 movie producers

Jan wants to know how long before a movie producer knocks. We want to convert one movie producer into weeks.

$$\frac{1 \text{ movie producer}}{1} \times \frac{250{,}000 \text{ weeks}}{430 \text{ movie producers}}$$

$\approx 581.395 \doteq 581$ weeks

936. How much pain is involved in finding the exact value of

$$\frac{78{,}398 \times 32\tfrac{1}{2} \times 0}{88888} \div \frac{5531.007 \times 39\tfrac{2}{3} \times 19}{44^4 - 33^7 + 1{,}000{,}005}$$

Not much pain at all. The numerator (the top) of the first fraction is equal to zero. The first fraction is equal to zero. Zero divided by any number is equal to zero.
    The whole thing is equal to zero.

939. $y^3 y^7 = ?$

    The rule is $x^m x^n = x^{m+n}$

    $y^3 y^7 = y^{10}$ The proof: $y^3 y^7 = (yyy)(yyyyyyy) = y^{10}$

| 949-951 | Complete Solutions and Answers |

949. You poured the 2 pounds, 11 ounces of Ralph's litter into the three litter boxes, dividing it evenly. How many ounces did each box receive?

    2 pounds, 11 ounces = 32 ounces + 11 ounces = 43 ounces

    43 ounces ÷ 3 boxes = 14 ⅓ ounces per box

950. Are you more likely to find anchovy ice cream or an eagle who can do calculus?

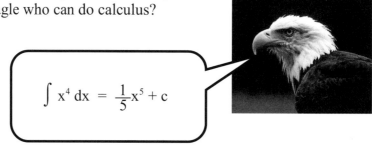

$$\int x^4 \, dx = \frac{1}{5}x^5 + c$$

    Stop the press! A reader put "anchovy ice cream" on an Internet search and found that it exists!
    Worse yet, there was a recipe for putting anchovy ice cream on a Caesar salad. I am speechless.

951. Let P be the set of all pizzas in the world right now. Find two sets, C and D, such that C ∪ D = P where C and D are *not* disjoint.

    Again, there are many possible answers, and yours may be different than mine.

My first answer: Let C be all the pizzas that weigh more than one pound. Let D be all the pizzas that weigh less than three pounds.

My second answer: Let C be all pizzas that contain either anchovies or olives or both. Let D be all pizzas that don't contain olives.

    That's a little trickier. C ∪ D will equal all the pizzas in the world, because any particular pizza will either contain olives or it won't. If it does, it is in C. If it doesn't, it will be in D.
    C and D are not disjoint, because every pizza that contains only anchovies will be in both sets.

My third answer: Let C and D both be the set of all pizzas in the world. Then they are not disjoint, because C = D. And C ∪ D will be the set of all pizzas in the world.

Complete Solutions and Answers | 955–962

955. She uses one twenty-fourth of a bar of soap for every 5 minutes she's showering. She spent an hour in the shower. How much soap did she use?

We'll work in minutes. We want to convert 60 minutes into bars of soap.

$$\frac{60 \text{ minutes}}{1} \times \frac{1/24 \text{ bars of soap}}{5 \text{ minutes}}$$

$= \frac{1}{2}$ bar of soap

$$\frac{60}{1} \times \frac{1}{24} \div 5 = \frac{60}{1} \times \frac{1}{24} \times \frac{1}{5} = \frac{1}{2}$$

960. Jan offered him 1½ cups of hot milk. Joe drank half of that. (His tummy was getting pretty full.) How much milk did Joe drink?

One-half of 1½ cups

$\frac{1}{2} \times 1\frac{1}{2}$

$\frac{1}{2} \times \frac{3}{2} = \frac{3}{4}$ He drank three-fourths of a cup of milk.

962. Junior took a nap from 12:55 to 4:30. How long did he sleep?

From 12:55 to 1:00   5 minutes
From 1:00 to 4:00   3 hours
From 4:00 to 4:30   30 minutes

Junior slept 3 hours and 35 minutes.

| 970 | Complete Solutions and Answers |

970. A can of Ralph's eyelashes has a volume of 50 cubic inches. One Ralph eyelash has a volume of $\frac{5}{134}$ cubic inches, how many of them are in a can?

There are two ways you might approach this problem.

First Way: Using the General Rule*, suppose that the can has a volume of 12 cubic inches and each eyelash had a volume of 4 cubic inches. Then there would be 3 of them in the can. You divided.

$$50 \div \frac{5}{134} = \frac{50}{1} \times \frac{134}{5} = \frac{\overset{10}{\cancel{50}}}{1} \times \frac{134}{\cancel{5}_{1}} = 1{,}340 \text{ eyelashes in the can.}$$

Second Way: Using a conversion factor. We know that one eyelash equals $\frac{5}{134}$ cubic inches. That gives us the conversion factor.

$$\frac{50 \text{ cubic inches}}{1} \times \frac{1 \text{ eyelash}}{(5/134) \text{ cubic inches}}$$

$$\frac{50 \;\cancel{\text{cubic inches}}}{1} \times \frac{1 \text{ eyelash}}{(5/134) \;\cancel{\text{cubic inches}}}$$

$$\frac{50 \text{ eyelashes}}{5/134}$$

$$50 \div \frac{5}{134} = \ldots = 1{,}340 \text{ eyelashes in the can.}$$

The three dots (...) are called an ellipsis. That means that something has been left out.

In this case I left out all the fraction stuff since I did all that in the First Way part of this problem.

---

* If you don't know whether to add, subtract, multiply or divide, first restate the problem with really simple numbers.

Complete Solutions and Answers | 977–989

977. Why can't you put a fake eyelash on your retina?

   For the same reason you can't put a fake eyelash on your kidney or your heart or your brain. Those things are inside your body.

   Your retinas are located at the back on your eyeballs. A retina changes light into electric signals which are send along the optic nerve to your brain.

980. Then Junior said something that made Janice's heart rate go from 80 to 180. What percent increase is that?

   An increase of 100 beats per minute. $(180 - 80)$
   100 is what percent of 80?
   $100 = ?\%$ of 80
   $100 \div 80 = 1.25 = 125\%$

989. Four bikes and three smart TVs cost $3955.
   Three bikes and two smart TVs cost $2810.
   How much does each cost?

   Let x = the cost of a bike
   Let y = the cost of a smart TV

   $$\begin{cases} 4x + 3y = 3955 \\ 3x + 2y = 2810 \end{cases}$$

Multiply the first equation by 3. Multiply the second equation by 4.

   $$\begin{cases} 12x + 9y = 11865 \\ 12x + 8y = 11240 \end{cases}$$

Subtract the bottom equation from the top.

   $y = 625$

Stick $y = 625$ into the first equation.

$4x + 3(625) = 3955$
$4x + 1875 = 3955$
$4x = 2080$
$x = 520$

Bikes cost $520 and smart TVs cost $625.

Index

.50¢
   #371.......... 39
   #690.......... 109

anchovies
   #950.......... 19

area
   #460.......... 32

business math
   #336.......... 101
   #452.......... 93
   #459.......... 14
   #532.......... 94
   #662.......... 17
   #829.......... 11
   #860.......... 13

cardinal numbers
   #629.......... 45
   #678.......... 103
   #733.......... 45
   #758.......... 90
   #837.......... 45
   #920.......... 65
   #930.......... 45

chemical equations—
balancing
   #162.......... 77
   #225.......... 77
   #333.......... 77
   #378.......... 79
   #469.......... 77
   #506.......... 77
   #600.......... 100
   #702.......... 78
   #714.......... 106

conversion factors
   #127.......... 30
   #137.......... 25
   #153.......... 37
   #204.......... 14
   #219.......... 25
   #234.......... 91
   #236.......... 30

#277.......... 16
#326.......... 18
#359.......... 108
#369.......... 14
#431.......... 35
#436.......... 68
#439.......... 96
#456.......... 16
#461.......... 46
#502.......... 74
#557.......... 36
#558.......... 12
#576.......... 65
#582.......... 113
#611.......... 111
#660.......... 14
#665.......... 21
#681.......... 72
#682.......... 75
#731.......... 19
#800.......... 57
#801.......... 67
#803.......... 69
#825.......... 26
#871.......... 76
#873.......... 104
#932.......... 76
#955.......... 106
#970.......... 55

d = rt
   #240.......... 99
   #359.......... 108
   #466.......... 49
   #502.......... 74
   #669.......... 65
   #773.......... 33
   #900.......... 31

decimals
   #134.......... 75
   #325.......... 32
   #666.......... 51
   #782.......... 72

divisibility rules
   #235.......... 84
   #475.......... 113
   #603.......... 88
   #711.......... 88

exponents
   #861.......... 19
   #863.......... 40

fractions
   #114.......... 9
   #119.......... 22
   #128.......... 12
   #135.......... 84
   #218.......... 9
   #251.......... 12
   #254.......... 22
   #321.......... 9
   #346.......... 84
   #409.......... 30
   #412.......... 9
   #470.......... 84
   #505.......... 97
   #565.......... 45
   #583.......... 105
   #595.......... 83
   #654.......... 26
   #656.......... 31
   #657.......... 26
   #667.......... 26
   #671.......... 85
   #725.......... 40
   #730.......... 29
   #734.......... 36
   #856.......... 67
   #857.......... 63
   #862.......... 55
   #918.......... 31
   #960.......... 63

Index

General Rule—
When to add, subtract, multiply, or divide
- #153.........37
- #224.........37
- #240.........99
- #370.........38
- #555.........75
- #601.........69
- #602.........94
- #652.........33
- #701.........69
- #880.........94
- #970.........55

genes—dominant and recessive
- #154.........54
- #253.........54
- #300.........105
- #328.........54
- #366.........99
- #413.........54
- #449.........105
- #467.........61
- #575.........54
- #668.........54

living/non-living
- #922.........11

moles
- #179.........79
- #227.........75
- #339.........75
- #485.........75
- #715.........94

negative numbers
- #221.........52
- #349.........52
- #457.........53
- #563.........53

parts of the body
- #112.........112
- #115.........95
- #131.........68
- #138.........101

- #157.........46
- #213.........101
- #226.........73
- #258.........70
- #261.........64
- #262.........59
- #275.........68
- #329.........46
- #337.........110
- #338.........73
- #360.........113
- #372.........68
- #425.........100
- #464.........59
- #508.........109
- #531.........90
- #626.........59
- #864.........57
- #977.........57

percents
- #121.........58
- #122.........48
- #129.........20
- #130.........35
- #132.........41
- #133.........52
- #136.........89
- #155.........60
- #166.........110
- #207.........14
- #209.........110
- #217.........20
- #241.........35
- #250.........39
- #252.........27
- #260.........107
- #276.........18
- #322.........27
- #324.........44
- #331.........25
- #348.........72
- #368.........96
- #410.........12
- #411.........38
- #437.........89
- #441.........74

- #455.........39
- #458.........22
- #462.........20
- #490.........110
- #501.........30
- #527.........21
- #541.........32
- #560.........9
- #561.........18
- #584.........111
- #593.........50
- #607.........103
- #653.........36
- #658.........9
- #676.........97
- #719.........97
- #722.........76
- #723.........15
- #724.........63
- #727.........11
- #728.........9
- #735.........42
- #754.........59
- #755.........47
- #756.........51
- #789.........88
- #790.........76
- #794.........78
- #802.........65
- #805.........72
- #806.........76
- #835.........57
- #858.........13
- #859.........17
- #865.........42
- #866.........51
- #891.........78
- #917.........26
- #921.........69
- #980.........114

photosynthesis
- #151.........43
- #160.........70
- #274.........46
- #433.........44

Index

pounds and ounces—
    feet and inches
  #106. 39
  #552. 69
  #562. 49
  #594. 78
  #628. 53
  #757. 61
  #759. 65
  #949. 40

receptors
  #222. 48
  #327. 49

sets
  #150. 10
  #176. 105
  #242. 105
  #255. 41
  #272. 10
  #323. 42
  #414. 25
  #432. 42
  #454. 10
  #525. 10
  #564. 38
  #577. 72
  #627. 42
  #659. 11
  #664. 13
  #687. 90
  #760. 55
  #824. 55
  #826. 24
  #828. 55
  #836. 33
  #901. 24
  #902. 34
  #951. 24

solving equations
  #156. 56
  #161. 66
  #206. 16
  #237. 66
  #257. 56
  #350. 87

  #373. 56
  #434. 56
  #529. 88
  #549. 66
  #579. 100
  #783. 74
  #827. 47
  #855. 19
  #870. 65
  #872. 88

terms in algebra expressions
  #158. 62
  #256. 62
  #347. 62

time lapse
  #152. 32
  #205. 27
  #377. 64
  #463. 27
  #465. 64
  #507. 102
  #526. 27
  #640. 47
  #655. 23
  #838. 90
  #962. 106

volume
  #220. 16
  #430. 18
  #645. 57
  #663. 18
  #774. 63
  #888. 61
  #919. 55

word problems
  #110. 81
  #123. 18
  #159. 64
  #163. 91
  #273. 43
  #279. 95
  #302. 22
  #320. 20
  #345. 66

  #374. 66
  #380. 66
  #500. 66
  #580. 56
  #596. 62
  #625. 67
  #639. 74
  #641. 78
  #646. 113
  #661. 28
  #680. 67
  #700. 83
  #721. 104
  #729. 24
  #732. 31
  #804. 90
  #830. 15

If you want to learn more about *Life of Fred* books visit

FredGauss.com